John Ellor Taylor

The Sagacity and Morality of Plants

A Sketch of the Life and Conduct of the Vegetable Kingdom

John Ellor Taylor

The Sagacity and Morality of Plants
A Sketch of the Life and Conduct of the Vegetable Kingdom

ISBN/EAN: 9783744772518

Printed in Europe, USA, Canada, Australia, Japan

Cover: Foto ©berggeist007 / pixelio.de

More available books at **www.hansebooks.com**

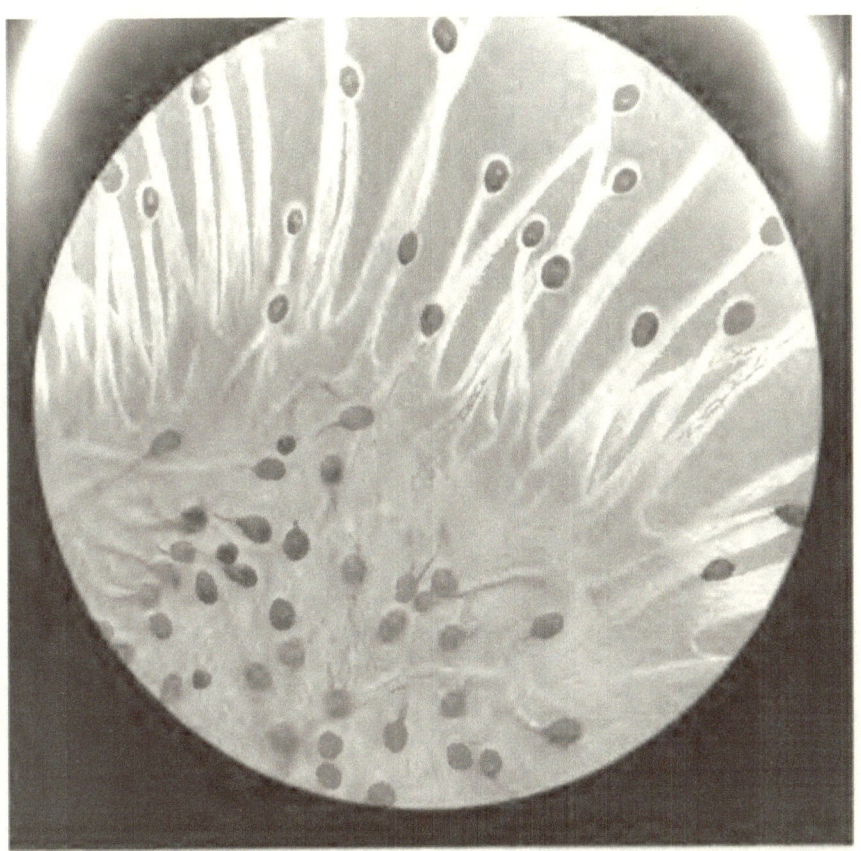

EDGE OF LEAF OF DROSERA ROTUNDIFOLIA.
× 30

THE SAGACITY & MORALITY OF PLANTS

A SKETCH OF THE
LIFE & CONDUCT OF THE VEGETABLE KINGDOM

BY

J. E. TAYLOR, Ph.D., F.L.S., F.G.S., ETC.

AUTHOR OF 'FLOWERS: THEIR SHAPES, PERFUMES, AND COLOURS,' ETC. ETC.
EDITOR OF 'SCIENCE-GOSSIP'

WITH COLOURED FRONTISPIECE AND 100 ILLUSTRATIONS

London
CHATTO AND WINDUS, PICCADILLY
1884

Printed by R. & R. CLARK, *Edinburgh.*

PREFACE.

THE reader may, if he so chooses, consider both the title of this Book, and much of its contents, as a Parable. But I have taken up the Parable with a view of bringing the Lives of Plants more nearly home to us. Botany is no longer a matter of counting stamens and pistils, and expressing the classified result in a Greek-derived nomenclature; it no longer consists in merely collecting as many kinds of plants as possible, whose dried and shrivelled remains are too often only the caricatures of their once living beauty. It is now a science of *Living Things*, and not of mechanical automata, and I have endeavoured to give my readers a glance at the laws of their lives. Therefore, whilst not beseeching criticism (seeing I have not written so much for learned botanists as for those who take an intelligent interest in plants), I do not deprecate it. Nobody is more conscious than the author that he has only lightly touched upon the fringe of a great subject; but if this little Book is the means of rendering plants and flowers more interesting to people after they have read it than they were before, it will not have been written in vain.

IPSWICH, *March 4, 1884.*

CONTENTS.

CHAPTER I.
		PAGE
Introduction	. . . .	1

CHAPTER II.
Misunderstandings	. . .	9

CHAPTER III.
Stating the Case	. . . .	17

CHAPTER IV.
Wood-Craft	. . . . .	38

CHAPTER V.
Floral Diplomacy	. . . .	56

CHAPTER VI.
Floral Diplomacy (*Continued*)	. .	70

CHAPTER VII.
HIDE AND SEEK 92

CHAPTER VIII.
"DEFENCE, NOT DEFIANCE" . . 115

CHAPTER IX.
CO-OPERATION . . . 160

CHAPTER X.
SOCIAL AND POLITICAL ECONOMY OF PLANTS . . 179

CHAPTER XI.
POVERTY AND BANKRUPTCY . . . 207

CHAPTER XII.
ROBBERY AND MURDER 226

CHAPTER XIII.
"TURNING THE TABLES" 257

CHAPTER XIV.
GEOGRAPHICAL VICISSITUDES OF PLANTS 285

LIST OF ILLUSTRATIONS.

	PAGE		PAGE
Allium vineale	187	*Centaurea cyanus*	171
Angelica sylvestris	172	Cherry, Umbel of	166
Annual Dog's Mercury	64	Common Comfrey	83
Arctostaphylos uva-ursii	97	Comparison of Plants, Zoophytes, and Insects	13
Arum	120		
Autumn Squill	186	Corn Bluebottle	171
		Crowberry	97
Barren Brome-grass	75	Cuckoo Pint	120
Bean-plant in first Growth	21		
Bean, Section of	18	Daisy, Section of	170
Bog-bean	146	Dandelion, showing Pappi	111
Bromus sterilis	75	Dead Nettle, Section of	84
Broom-rape	250	Dead Nettle, White	86
Butcher's Broom, The	42	Dodder, The	246
		Drosera anglica, Leaf of	261
Cabbage-leaf, Malformation of	270	*Drosera obovata*, Leaf of	261
		Drosera rotundifolia	260
Carnation, showing Pistil	77	*Drosera rotundifolia*, Glands of	263
Cells of Potato, showing Starch-grains	23		
Cells of *Tradescantia*	8	Embryo of Bean	20

LIST OF ILLUSTRATIONS.

	PAGE		PAGE
Empetrum nigrum	97	*Menyanthes trifoliata*	146
Euphorbia portlandica	201	*Mercurialis annua*	64
		Mistletoe, Flower of	238
Fig, Section of	100	Mistletoe, Fruit and Seed	240
Filago canescens	147	Mistletoe, Sections of	239
Filago spathulata	147		
Flower of Vine	33	Nepenthes	276
Fool's Parsley, Umbel of	166		
Fossil Fungus	253	Orchid, Flower of	87
Frog's-mouth, Flower of	87	Orchid, Ladies' Tresses	90
Fungus on Leaf	255	Orchid, Pollinia of	88
		Orchid, Section of	88
Garlic	187	*Orchis militaris*	89
Geranium Robertianum	148	*Orchis purpurea*	89
Glumes of Wall Barley	132	*Orchis simia*	89
Gooseberry Leaves and Scales	182	*Orobanche rapum*	250
		Oxalis acetosella	125
Grass of Parnassus	284	Ox-eye Daisy	169
Grass Pea	204		
		Parietaria officinalis	65
Herb-Robert	148	*Parnassia palustris*	284
Hordeum murinum	131	Pellitory of the Wall, The	65
		Pistil of Flower, magnified	36
Juncus bufonius	209	Pistil of Flower showing Pollen-tubes	37
Lamium album	86		
Lathræa squamaria	251	Pistillate Flowers of Oak	72
Lathyrus aphaca	52	Pitcher-plant	276
Lathyrus nissolia	204	Pollen-grains, magnified	35
Leaf of Pea	55	Portland Spurge, The	201
Loranthus europæus	242	Primrose, Fertilisation of	79
Lythrum salicaria	80	Primrose, Pin Centre	78

LIST OF ILLUSTRATIONS.

	PAGE
Primrose, Rose Centre	78
Purple Loosestrife	80
Ranunculus aquatilis	197
Ranunculus drouettii	199
Ranunculus heterophyllus	198
Ranunculus tripartus	199
Rose Fruit	101
Rubus cæsius, Flower-bud of	151
Rubus cæsius, Stem of	150
Rubus fusco-ater, Flower-bud of	151
Rubus glandulosus, Stem of	150
Rubus nemorosus, Flower-bud of	151
Rubus rhamnifolius, Stem of	150
Rubus rudis, Stem of	150
Rubus umbrosus, Flower-bud of	151
Ruscus aculeatus	42
Salvia, Fertilisation of	85
Salvia, Flower and Stamens of	84
Sarracenia, Pitchers, etc., of	273
Scarlet Bearberry	97
Scilla autumnale	186
Scrophularia, Flower of	87
Section of Leaf, showing Cells and *Chlorophyll*	29
Section of Root-tip of Plant	25
Spiranthes autumnalis	90
Staminate Flowers of Oak	73
Stomata of Leaf	27
Strawberry	100
Structure of Wallflower	79
Sundew, The	260
Symphytum officinale	83
Tankard Turnip	188
Tendrils of Vine	55
Toad Rush, The	209
Toothwort, The	251
Tubers of *Orchis mascula*	184
Turnip, Tap-root of	188
Under-surface of Leaf, showing *Stomata*	27
Venus' Fly-trap, Leaf of	267
Venus' Fly-trap, The	266
Vine, Flower of	33
Wall Barley	131
Water Crowfoot	197
Wood Sorrel	125
Yellow Vetch, The	52

THE SAGACITY
AND
MORALITY OF PLANTS.

CHAPTER I.

INTRODUCTION.

I SHALL be met at the outset with the remark that the term Morality can only be rightly applied to conscious agents, and that plants cannot be classed among that number. Has this commonly adopted opinion been proved, or do we accept it as a traditional conclusion which to most people seems self-evident? Wordsworth did not think so. He said—

> " It is my faith that every flower which blows
> Enjoys the air it breathes ! "

Darwin, speaking of the sensitiveness of the root-tips of plants, shows they have acquired diverse kinds of sensitiveness, and that " it is hardly an exaggeration to say that the tip of the radicle thus endowed, and having the power of directing the movements of the adjoining parts, acts like the brain of one of the lower animals; the brain being seated within the anterior end of the body, receiving impressions from the sense-organs, and directing the several movements."

It is only within the last few years, since botany has been studied from its biological side, that we have wakened up to understand what wonderful objects plants are. A new language has been developed in which to describe their novel relationships. Whether we believe in the consciousness of plant-life or not, this language almost implies such a belief. We speak of plants adopting this habit or that device —always and only when such habits and devices are beneficial to them—as if they did it of set and intelligent purpose. The works of Darwin, Lubbock, Müller, Wallace, Kerner, Grant Allen, Wilson, and others, are particularly noticeable for this style of description. Who knows—perhaps there can be no life, animal or vegetable, unaccompanied by consciousness! The minutest animalcule, lowest placed in the scale of animal being, displays a consciousness of external surroundings as simple and elementary as its own structure; and its life is passed in voluntarily and automatically responding to them. From this vanishing point of intelligence up to the genius of Shakespere, if we were in possession of all the facts, it might be possible to intercalate every intermediate stage of mental development and difference.

Perhaps one reason why plants have been denied

consciousness and intelligence is because in the structure of even the highest developed species we find no specialised nerves or tracks along which sensations can travel, or where they can be registered, as in the ganglia and brains of the higher animals. But it should be remembered that none of the creatures included in the ancient and widely distributed sub-kingdom *Protozoa* possess a nervous structure, whilst many in the next more highly organised sub-kingdom *Cœlenterata* have no trace, and in the rest but a feeble development. Yet we do not deny these lowly organised animals a dim and diffused consciousness, or even the possibility of their structures being so modified that they can profit by experience, and thus develop that accumulated experience of their kind we call instinct.

The physical basis of life, Protoplasm, is the same for plants as for animals. The first differentiated or modified form of this we meet with is the curious animalcule called *Amœba*. As we watch its movements we cannot refrain from ascribing to it some dim consciousness of the life it leads. But amœboid structure is common even in the lowest kinds of plants, and amœboid movements can be seen in some of their tissues. Witness also the habits and intelligent movements of the zoospores of seaweeds and many other Algæ, and the locomotion of the antherozoa of Mosses, Ferns, etc. Not many years ago these objects were classed as animalcules on this account, and nobody then doubted these so-called little animals behaved consciously and intelligently.

The life of a plant, like that of an animal, is a series of constant adjustments between internal structure and organisation and external surroundings. The latter are of such an almost infinitely variable and varying character that we cannot wonder the adjustments, or in other words the *habits*, of plants are so infinitely numerous—especially when we remember the long geological periods of time during which the constant adaptations and modifications have been taking place.

We sometimes hear people speak of the "instinct" of plants. But how can instincts arise unless there be some kind of consciousness? For instinct is now generally regarded as the experience of the *race*, as distinguished from that of the *individual*. The registration of experience may be in itself an intelligent rather than an automatic act. To speak of the "tendencies" on the part of plants to assume a certain habit is merely to coin a word to disguise our ignorance of the process. A "tendency" for a plant to behave in a certain manner is merely a habit based upon the past experience of its ancestors as to what has proved best for them as a species.

Many people still regard even the higher animals as automata. They are unwilling to allow that the various intelligent acts they perform proceed from cerebration, exactly in the same way as the intelligent and rational actions of men. They formulate the existence of a different kind of mentality, which they frequently call "instinct," "sagacity," etc. But this old-fashioned way of disposing of the psychology of the lower animals will soon become extinct. Comparative animal psychology is rising to the dignity of a special science, and it has already fairly demonstrated that the intelligence of the dog, horse, and elephant, differs mainly in degree, and not in kind, from that displayed by man.

If animals have been regarded as mere automata —as living machines moved from without—we cannot be surprised that plants should still be unquestionably placed in that category. To speak of *Vegetable Psychology* would cause a smile to ripple over the faces even of those who have granted the identity of the intelligence between man and the brute. But the near future may have occasion to show there can be no life absolutely without *psychological* action—that the latter is the result of the former. It may some day be shown that life is conditioned by psychological action; and that there is in plants the equivalent of "instinct" in animals —the power of gaining individual experience, and of transferring such experience to descendants to profit thereby, not altogether unconsciously!

It is true that natural selection, by weeding out of existence all plants which cannot adapt themselves to their environment, sternly insists upon the adaptation of the survivors. It may be observed that the various characters and habits of plants have been produced by natural selection. To this it may be answered that so also have the characters and habits of all animals, including man; and, nevertheless, "instinct" and "sagacity" are allowed them, in spite of (rather, perhaps, because of) natural selection.

Nothing can be more marked, even among animals, than the likes and dislikes of plants. Human beings can hardly express the same feelings more decidedly. Some species prefer the light, others the shade. Some will only live in arid deserts, others in swampy marshes and morasses. There are plants which love the heat like a Hindoo, and other kinds which revel amid the snows of Arctic regions or Alpine slopes. Some are soda-lovers as regards their mineral dietary, and therefore flourish best where that element is present, as by the seacoast, on hillsides where soda-felspar decomposes, or by the runnels of Cheshire and Worcestershire brine-springs. We have plants which grow most luxuriously where lime is abundant, as the clovers; and others which intensely dislike it, as the Heather and Foxglove. Nay, there is, perhaps, even a "messmateship" among plants, which inclines species to prefer to grow in company, as the Yellow-wort (*Chlora perfoliata*) seems to do with the Bee Orchis (*Ophrys apifera*).

Hosts of common plants constantly perform actions which, if they were done by human beings, would at once be brought within the category of right and wrong. There is hardly a virtue or a vice which has not its counterpart in the actions of the vegetable kingdom. As regards conduct, in this respect, there is small difference between the lower animals and plants.

Responsibility is another matter. It can only be attained when consciousness and volition have reached a certain stage of development. Volition, as Professor Cope has recently shown (*On the Evolutional Significance of Human Character*), is the latest evolved of mental characteristics. Praise and blame, reward and punishment, can only be applied when agents have attained the higher stage of conciousness. Nobody thinks of condemning infants or idiots, and our censure of children and animals is of the mildest kind. The conduct of plants falls still further outside any tribunal of this kind. But it is none the less interesting to find that, as regards the behaviour of vegetable organisms, in common with all other living beings,

"One touch of Nature makes the whole world kin!"

No botanist, working with the microscope, who has watched the streams of protoplasm ebbing and flowing within the cell, or from vessel to vessel, can feel that plants are the inert and lowly organised objects popular opinion unquestionedly holds them to be. The physiological work constantly going on even in the humblest of vegetable forms is of a highly complex physical and chemical character. Whether life be the cause or the product, the process is none the less wonderful. Before the revelations which the microscope gives us of the inner life of the minutest plants, the boldest researches of science stand checked. What is above and beyond these microscopic protoplasmic streams? We stand on the verge of the unknown even when beholding the living changes taking place within the glassy frustule of a diatom!

FIG. 1.—Cell of *Tradescantia*, drawn at intervals, and showing changes in the contained protoplasm.

> "And what if all of animated Nature
> Be but organic harps, divinely framed,
> And trembling into thought, as o'er them sweeps
> Plastic and vast, one universal breeze;
> At once the soul of each and God of all?"

CHAPTER II.

MISUNDERSTANDINGS.

PROBABLY we have acquired a more intimate and genuine knowledge of the life-histories of plants, within the last fifteen years than in the whole preceding history of botanical science. In this knowledge the information respecting the devices put forth by flowers to ensure cross-fertilisation, are perhaps both the most interesting and the most incomplete. Incomplete, not because the species of flowers which have been carefully observed have disappointed us in the results, but because so many additional observations are required on numerous other kinds. Here lies one of the charms of modern botany—the humblest observer, by confining himself to one or two species of flowers, whose devices for securing insect services in procuring cross-fertilisation are as yet unknown or only partially known, may render loyal and genuine service to the cause of science. In no department of research does careful observation go for more than in botany.

It is only within the brief period referred to that the importance of crossing, in flowers, has been fully understood. Variety is the charm of life, from a moss to a man. Interchange of conditions are beneficial, unless they are extreme. Most farmers know that the wheat grown from seed raised on the same field is more liable to "smut" than if it had been raised from seed grown on another farm. Grow the same crop on the same field, and in a very few years the land will "sicken," and refuse to bear. An interchange in those living parts of plants detached for reproductive purposes helps to secure strong plants and abundance of seeds, which are more likely to come off conquerors in that keen and never-ending Battle of Life which is going on in every green lane and field, and along every hillside.

We have learned to approach the study of plants from more than one new side. Flowers are no longer regarded as created solely for the uses and pleasure of man. Every character and quality they possess has been acquired by them and their ancestry in the struggle for existence. Every distinction individualising each species has come into existence because it has proved best for its well-being. We cannot wonder, therefore, that there are few species of plants which have not a museum of their own—a series of botanical curiosities, such as aborted organs or higher-developed ones, tendencies to "sport" in a certain kind of way, etc.,—all of which, by the sidelights thrown upon them by the doctrine of evolution, now assist the botanist in working out the ancestral history of plants in a marvellous and unexpected manner.

The real nature of a plant is hardly understood, even by many who have taken up botany as a holiday science. We hear so much of *flowers*, of their elegant shapes, their delicately tinted or gorgeously coloured petals, and of their delightful perfumes, that many people imagine plants must have been formed for the special purpose of bearing flowers alone! To such people it seems almost libellous to say that every flower is in reality composed of aborted and degraded leaves—that a leaf is a much more highly-organised vegetable production than a petal! Yet such a statement is quite correct. Leaf-buds can be changed at will into flower-buds, as every gardener knows. This transformation is effected by *starving* and crippling the plant, not by feeding it. The horticulturist "rings," or otherwise cripples his Azaleas and Camellias, so as to force them to produce flower-buds instead of leaf-buds. In other words, he decreases the food supply, and so produces a part less highly organised than leaves would be.

In nature the same end is effected in other ways. Flowers are frequently terminal—that is to say, are borne at the ends of branches, where the uprising sap or fluid food must be thinnest and poorest, seeing it has been tapped all the way up by the leaves. Hence these terminal starved buds tend to develop into flowers. Flowers are also produced in the axils of leaves, or on separate long stalks, as the Daisy and Dandelion, so that a large quantity of sap cannot reach them. If too much of food supply should be furnished, the latter kind of flowers "sport," and "monstrosities" are invariably the result.

A plant or tree, small or great, is in reality a colony of vegetable organisms—just as a "Sea-fir" (*Sertularia*) among the Zoophytes is a colony of hydra-like animals. The latter have a fluid flesh (*sarcode*) in common, just as leaves and flowers are supplied by the same flow of sap. At certain times of the year the Sertularians develop special buds or capsules (*gonophores*), which contain the larvæ or young. These are cast in the water, and usually assume some jelly-fish-like shape. This parallelism between a colony of Sertularian polypes, vegetative and reproductive, and a plant with its leaf-buds and flower-buds may be still further illustrated by the life-history of the common *Aphis*. Generation after generation is produced by the latter, without the intervention of the male insect, by a process of budding. Then ensue the sexual individuals, possessing wings, so that they can remove to a distance, and commence the work of founding a new Aphis colony. Similarly the parts of plants we call flowers are produced for the special purpose of propagating the species. Thus every individual part of plants contributes somewhat to the well-

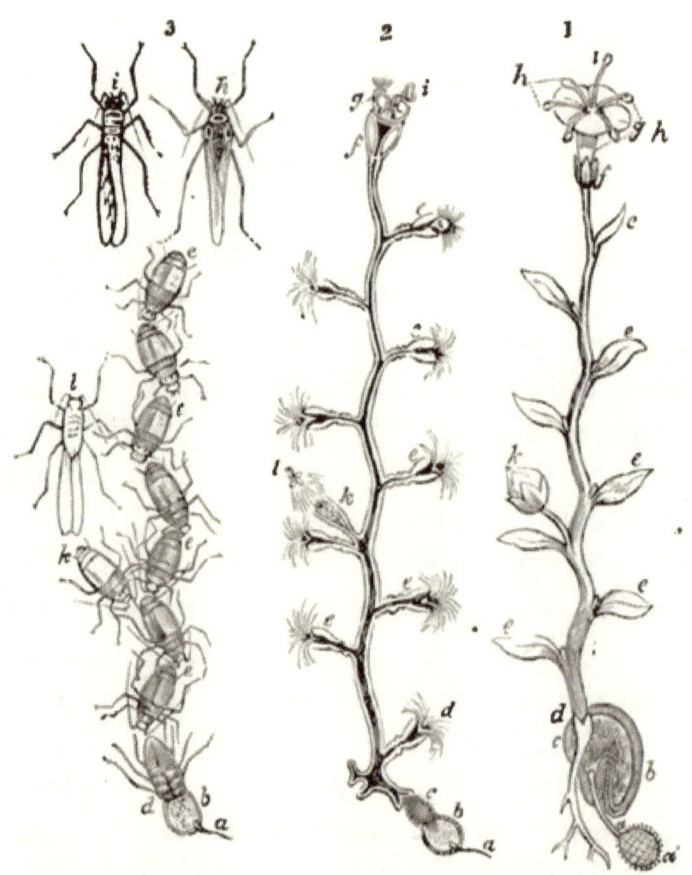

FIG. 2.—Comparison of Development in (1) a Flowering Plant ; (2) a Zoophyte ; and (3) a Colony of Plant-Lice (*Aphides*).—After Wilson.

being or perpetuation of the species to which it belongs.

A plant, therefore, is a Colony of Leaves, all of

which exist together on the Sertularian or co-operative principle. The flowers are simply leaf-organs naturally aborted and modified to serve a use altogether outside the individual wellbeing of the plant which produces them. Nay, in many instances, as we shall see further on, flowers are often produced at a great physiological expense, and with much individual loss to the parent plant.

Moreover, leaves and flowers exercise quite different functions in plant economy. The former accumulate energy; the latter expend it. Leaves assimilate carbon and liberate oxygen; flowers actually require oxygen, and liberate carbonic acid. In the latter respect they behave as if they were animals. Leaves store up food materials, either in bulb, corm, tuber, rhizome, or stem. Flowers exhaust this banking account by drawing upon it. Hence, we have only to supply the Hyacinth bulb with water, and it develops in the glasses placed for the purpose in our window, and brings forth both green leaves and the spike of coloured and sweetly-perfumed flowers. All these have been mysteriously metamorphosed out of the food-stuffs contained in the dry and seemingly quite inert bulb we purchased for sixpence.

Now we can understand the fact, variously interpreted and usually mis-stated, that some plants live for years before producing flowers. The Aloe, according to popular belief, flowers only once in a century! Some species of Alpine Gentians produce leaves for several years before they bear any flowers. At the end of this leafing season, and as the reward for such a persistent accumulation over expenditure, there is a decent banking account stored up usually in the shape of a root-stock or underground rhizome. Then follows the season of extravagance. The gorgeous development of flowers draws on every molecule thus prepared for it, keeps open house for insects and even birds, produces stores of seeds rich in a nursery supply of nitrogen and phosphorus; in short beggars the plant, and either ruins it altogether, or reduces it to a state of temporary bankruptcy, from which it can only recover by perhaps years of subsequent thrift! Every horticulturist knows that an abundant year of apples and pears is usually followed by one of dearth. In short, the "seven years of famine" are in this way contingent upon the "seven years of plenty."

The desire "to found a family" is as manifest among plants as among men! Otherwise what means this slow accumulation of energy on their part? All is eventually expended in the production of flowers and seeds. On the food-store possessed by the latter largely depends their chances of success in the battle of life. Consequently there is as great a "tendency" to accumulate with plants as with ourselves. "No man liveth to himself alone" (unless he wishes to be perpetually branded for selfishness), and this principle of altruistic morality applies to the vegetable kingdom as well. If I am allowed to use the term "morality" when speaking of the behaviour of plants, I should say the latter instinctively obey the same code, and none other, as that enforced upon humanity!

CHAPTER III.

STATING THE CASE.

A LAWYER does not think he is insulting the common sense or learning of the judge before whom he pleads, by stating his case to the jury in terms of legal explanation which for years may have been the A B C of his lordship. And no genuine botanist will quarrel with a writer who adopts the same plan with intelligent readers unpossessed of a scientific knowledge of plants.

The present chapter, therefore, will be devoted to a brief statement of plant-life as a whole. This is necessary to the general reader who desires to understand the full scope of the argument hereafter used.

There may be a few people who think the process of plant growth and development is less wonderful than it was, because more is known about it. I suppose there is always a certain condition of mind in which familiarity is sure to breed contempt. But, in my opinion, increased knowledge brings increased wonder. Let a man take a grain of wheat, just as it is, and keep it in some dry, cold place. No change occurs; but if he places it in a damp, warm soil, by and by he beholds a marvellous transformation! The plant sprouts and grows. How is this?

Until lately people regarded *growth* as a mystical and mysterious process altogether, and so saved themselves the necessity of any further explanation. Others invented *phrases* to account for it (not a bad method in philosophy), and such people wisely said that *growth* was due to the *vitality* of the seed or the *virtues* of the soil. The latter was a powerful affirmative, from which only a dejected few dared to turn away unsatisfied.

Let us see, in a quiet and matter-of-fact way, what actually does occur when a seed is placed in the ground.

Perhaps a botanist would notice, first of all, whether such seed had one lobe or two. If we take off the outer skin of a bean or acorn, we see two halves or *cotyledons*. No such appearance, however, is visible in a grain of wheat, barley, or maize, for these are seeds with one lobe only. The broadest classification of flowering-plants is based on this seed-difference—the two-lobed being termed

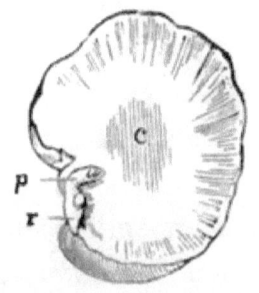

FIG. 3.—Bean in Section. *c*, one of the cotyledons; *p*, young stem; *r*, the young root.

dicotyledonous, and the single-lobed seeds *monocotyledonous*.

The sizes of seeds has little or no reference to the magnitude the plant will attain unto which sprouts from it. The Oak-tree does not bear larger seeds than the Bean-plant; those of the Pine and Poplar are not so large as those of the Pea.

Why does any variation occur at all? Everybody knows that most seeds are eatable, such as beans, peas, maize, wheat, hazel-nuts, walnuts, etc. In most of them their nutritious matter is improved by cooking. Even in those seeds we cannot eat—such as the acorn, horse-chestnut, etc.—it is not because nutritious matter is absent, but because of the presence of some bitter, disagreeable, or even poisonous substance diffused through the nutritious store. We shall see presently that these bitter or poisonous principles are protective to the seeds, and prevent their being so completely eaten up by hungry animals that none remain to perpetuate the species.

If we examine a common broad bean, split open, we see a greenish-white object called the *embryo*. This is the undeveloped plant which will sprout from the seed under favourable conditions. It is for this insignificant-looking object that, in reality, all the store of nutritious matter in the two lobes has been accumulated and stored away! If we planted such a bean in the soil, the first thing it would do would be to swell from the moisture it absorbed. The skin which covered it would then burst, the two seed-lobes would part asunder, but they would not separate, for the delicate embryo has a firm hold upon them. These two halves now act as a "wet-nurse" for the embryo — it draws all its nourishment from them for a time, just as an infant abstracts its food from the breasts of its mother. As it feeds on what the cotyledons contain it gets stronger and increases in size. One part of the embryo (the *radicle*) slowly, and by an instinct which never fails, makes its way into the soil and becomes the root. The other part (the *plumule*) just as undeviatingly grows upwards, not unfrequently carrying the cotyledons with it, until they are lifted quite out of the soil where we planted them. Henceforth the vegetable infant will get all the nourishment it requires, and more, from the soil and the atmosphere.

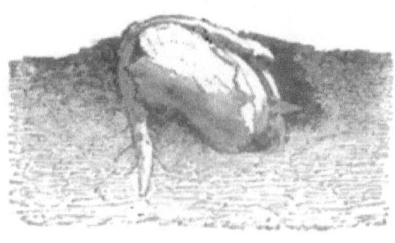

FIG. 4.—Embryo of Bean beginning to sprout in the Soil.

Growth in plants is produced by the multiplication of cells. If a man were to place a single brick in a bit of freehold land, and expect it to grow into a house, we should think he showed undoubted fitness for a lunatic asylum. Nevertheless, the actual growth of every plant is as if one brick of a house had the magical power of forming others. No fact is better

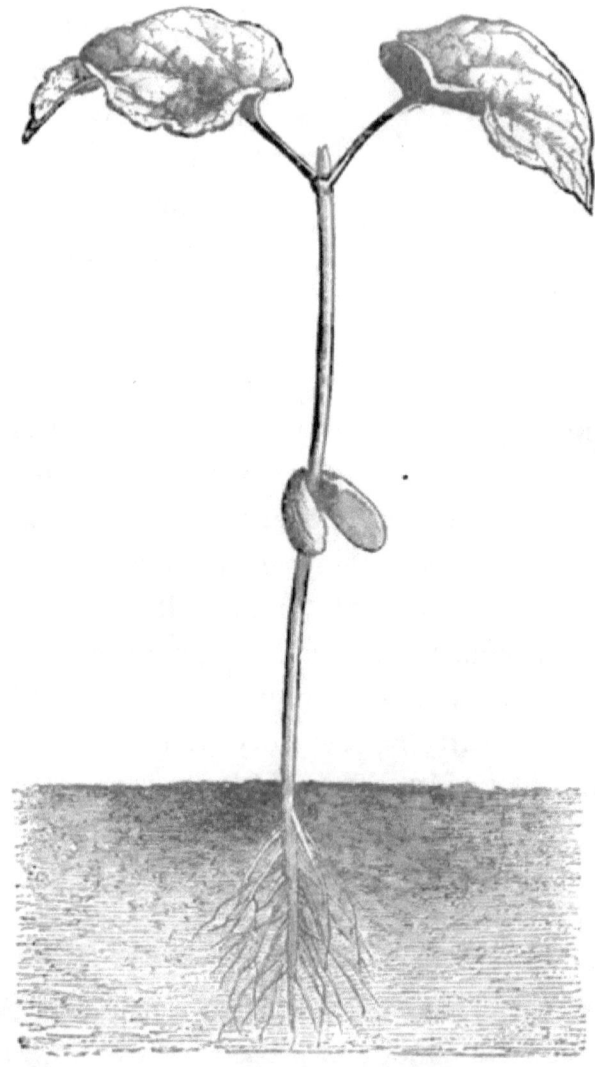

FIG. 5.—Bean-plant in its first Growth. The seed-lobes have been lifted out of the ground by the growth of the stem.

known than that if we prepare a thin slice of a plant

and look at it through the microscope, we shall see cellular and vascular structure—in other words, both cells and vessels packed together. When the young embryo-plant quickens into activity by the moisture and heat of the soil, its cell-growth is stimulated. Each cell produces another cell, by the simple method of self-division. Two cells are thus formed, which immediately produce two more, and thus in increasing multiple proportion they accumulate, until in a short space of time the original cell is perhaps the parent of hundreds of millions. Increase of bulk must of necessity take place when the cells are multiplied so enormously.

What delicate things these cells are, and what numerous shapes they assume! The most natural form is an oval, but they increase so fast and press so much against each other during growth that they are often formed with many sides. Then, again, how varied are the contents of these cells! They may be regarded as so many organic chemical laboratories, in which synthesis is carried on even more vigorously than analysis. Some are starch manufacturers like Colman, as in the Potato and other tubers and bulbs; some are perfume distillers like Rimmel, as the cells in the leaves of the Sweet-briar, Lavender, and Mints. Every cluster of cells and vessels has a work to do—sometimes special kinds of work, but usually generalised kinds. In the leaves of nearly all plants the cells are filled with green grains called *Chlorophyll*, and these are chemical laboratories in which the most important work in the whole organic world is carried on. We shall see more about them presently.

In the upward process of the growth of plants many of the cells change their shapes, and also transform their cell contents. The cells get drawn out into long, tapering vessels which are packed against each other as if they were spliced. In this way the tissue becomes very strong. It is converted into true wood by the internal formation of a substance called *lignine*. In herbaceous plants, however, the latter is seldom formed except in the veins, etc., of leaves, hence the reason why the latter can be "skeletonised." The cells of the roots of plants seldom secrete *chlorophyll*; their sap, instead, is usually of an *acid* character, especially in the fine rootlets.

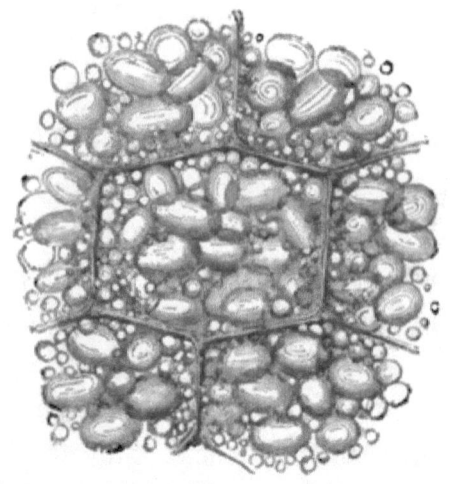

FIG. 6.—Cells of a Potato magnified, showing Starch-grains in their Interiors.

The cells of plants are as particular about the material used in their construction as a brickmaker who is anxious to supply really good bricks is to the kind of clay he uses. Only certain materials will do for the cells, and if the soil does not contain these, growth or cell-multiplication cannot take place. Moreover, the cells which build up the structures of the various parts of a plant often require different sorts of materials. Some of these are taken from the atmosphere (the *carbon* entirely so), and others from the soil. The food-material in the soil, however, is useless until it is dissolved in such a way that the cells can absorb it. Hence, if a plant be kept without water it will die of starvation even in the midst of plenty. Water dissolves most of the mineral salts out of the soil; these are absorbed by the roots, and employed to build up the tissues of the plant. The function of the roots is to extract such mineral salts out of the soil, whilst that of the leaves is to obtain carbon and oxygen from the atmosphere.

The tip of the root of any plant is a most wonderful object. If we examine a carefully prepared slice of it with a high magnifying power, we not only see the cells which build it up, but also a special set at the tip, forming a delicate sheath. When a root is growing and extending itself through the soil it comes into contact with stones, pebbles, and large grains of sand, all of which tend not only to obstruct, but also to injure it. It is to protect the *real* tip of the root—whose duty is to find the mineral salts diffused through the soil, and to abstract them—that the special layer of cells composing the *sheath* is formed over it. As fast as the latter are worn away, fresh ones take their places. This goes on until the root has penetrated as far as is necessary. The sheath-cells then die, for their special work is done; and now the fresh, uninjured cells of the root begin to collect and dissolve the

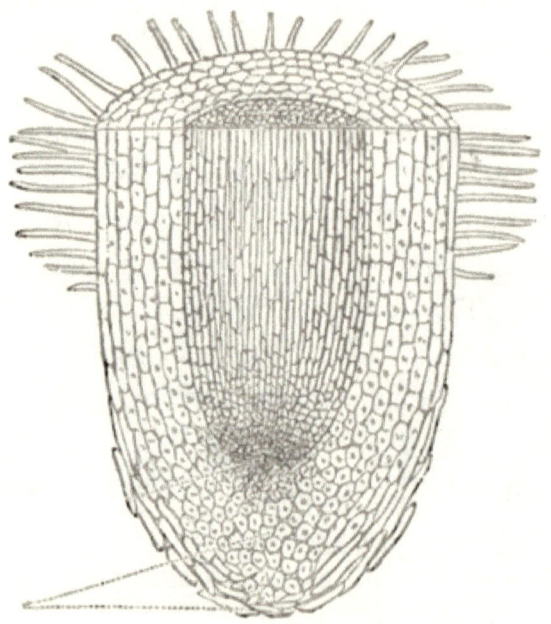

FIG. 7.—Section of the Root-tip of a Plant, showing the Sheath at the end, which protects the Plant.

mineral salts. The acid nature of the sap assists in this process. The dissolved salts pass from one cell to another, through the intermediate walls, as if they were strained through. We hardly need say that this upflowing fluid is called *sap*, and that it is to plants what the arterial blood is to animals.

But in an animal's body (at least in the higher kinds) the blood is sent to every part by means of that special muscular organ, the heart. In a plant or tree, no matter how large it is, no special arrangement of this kind exists for pumping the nutritious fluid. Nevertheless, the *sap*, or vegetable blood, finds its way to the extremest tip of the farthest twig!

Nature is seldom embarrassed for want of a device. Indeed, her extremity is usually her best opportunity. By means of "capillary attraction" the identical method which causes water to be "sucked up" by a dry sponge, or to permeate a mass of dry loaf-sugar, the sap in plants rises and passes upwards and onwards through the entire tissues of every plant. This upward flowing is much helped by the leaves, which in the daytime imperceptibly give off a great deal of water. In this way the leaves of an ordinary-sized Sunflower will transpire as much water as a man perspires. Thus the moisture containing mineral salts is always entering in at the roots, whilst it is evaporating at the leaves. It is like drawing water at a tap which cannot be drawn off there without the entire volume of water in the pipe being on the move. Even the wind, by rocking trees and plants to and fro, helps to pump the sap upwards. By one or other or all of these means the nutritious fluid of plants is as certainly diffused to every part as if each possessed a heart for the special purpose of supplying it!

The leaves of plants, however, are by far the busiest and most highly-organised of all their parts. They are built up of cells like other tissues; but these cells vary both in shape and size. If the under skin of almost any leaf be peeled off and examined under the microscope, we see not only the sinuous outline of each cell, but an immense number of mouths, or *stomata*, each provided with movable lips. Every square inch of the under surface of a lilac-leaf has nearly a quarter of a million such mouths. If we cut a leaf in section across one of these *stomata*, under the microscope

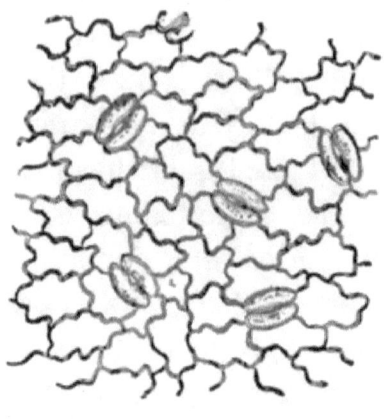

FIG. 8.—Under-surface of a Leaf (magnified), showing the *Stomata*, or Breathing-mouths.

FIG. 9.—Magnified Section of a Leaf, cut across a Breathing-mouth, to show the Structure of the latter.

we are able to see the complete mechanism of the structure, and also how atmospheric air gets into the interior of leaf-tissue. The openings into this tissue are rightly called "mouths," for they are actually the chief feeding-places of a plant. It is by their means all the carbon is obtained which afterwards is transformed into the wood of the tree or shrub, the starch of the potato or bulb, or the sugary matter of the beet, carrot, parsnip, and sugar-cane, and honey of the nectaries of flowers.

Let us now return to consider the nature and function of the *chlorophyll*, the chief substance secreted in the leaves of plants. An enlarged section of a leaf shows us the cells packed away, each containing the minute green grains of this important material. It is to the presence of *chlorophyll* in the cells that the greenness of all vegetation is due.

Everybody knows that the air we breathe is tainted with carbonic acid, cast into the atmosphere from the breath of the untold millions of men and animals inhabiting the globe. It has been calculated that at least forty-five millions of tons of carbonic acid are thus thrown into the atmosphere every day ! This has been going on during all geological time ; it will continue until the "last man" and the last animal has passed away into extinction !

Evidently, if there were no means of getting rid of this poisonous gas, the atmosphere would very speedily be unfit for animals to breathe. Here it is that the services of plants are conspicuous, apart from the food they supply to the animal world. In the daytime, and especially when the sun is shining, the myriads of *stomata* on the under-surfaces of leaves are all at work. The *stomata* are then wide open, and the atmospheric air is free to enter therein, carrying with it the carbonic acid gas which pollutes it. The air circulates in and out of the loose cellular interspaces. As soon as the carbonic acid it carries comes into contact with the *chlorophyll* or "leaf-green," a magic chemical change ensues. Carbonic acid is a molecule, or cluster of three atoms, two of which are oxygen, and the single one carbon. These are temporarily connected in a very remarkable manner. But the *chlorophyll* has the power of unpicking this cluster of three atoms, so as to get at the carbon it wants. The carbon is seized upon and absorbed, and the oxygen atoms are then turned loose into the atmosphere again. That very carbon atom

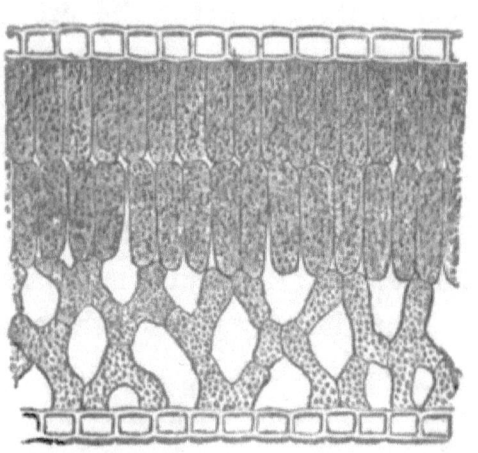

FIG. 10.—Magnified Section of a Leaf, showing the Cells within, full of *Chlorophyll*, or Green-colouring Matter.

was perhaps once in our blood. It was set free at the lungs, where it combined with the oxygen, and thus became part of the carbonic acid. Now the same carbon atom has been absorbed by the "leaf-green." By and by it will be transformed into sugar, starch, wood, or some other vegetable substance.

The leaves of plants are all arranged on the stems and branches so that each can do its work without interfering with the rest. Their multitudinous shapes, also, have distinct reference to their capability of decomposing carbonic acid. The process of spiral growth (*phyllotaxy*) gradually pushes out the leaves we see packed away in every leaf-bud, until at the end of the summer a long twig, with leaves arranged along all its length, has developed from it. The places where the leaves have developed are called *nodes*—the spaces between the leaves are known as *inter-nodal spaces*. The characteristic foliage of a tree very much depends on the *length* of the spaces between the leaves. Sometimes we have no spaces at all—the latter are "suppressed." Then we have the rosette arrangement of leaves matting the ground like those of the Daisy, Thistle, Hawkweed, and many others. When the leaves are large, as in the Horse-chestnut, there are few of them composing the spiral; when the leaves are small they are numerous, as in the Pine and Larch-trees. Usually, the inter-nodal spaces get shorter towards the end of a branch, and sometimes they are almost suppressed there, so that the terminal leaves are arranged like a rosette. This is the case with the Rhododendron and the Poinsettia. In the latter plant the rosette-grouped leaves are of a bright scarlet colour, and so attractive that the flowers are placed there, and no gorgeous petals are required to attract insects. Those desirous of seeing how the leaves of plants owe their shapes to their ability to obtain carbonic acid in their natural habitats, should read Grant Allen's charming work on *The Shapes of Leaves.*

The leaves of plants have received their characteristic shapes and positions under circumstances where *chance* and not *law* seems to the thoughtless mind the most probable agent. The leaves are fixed, and cannot go in search of the carbon atoms. The wind, however, brings the latter to them, and every gentle breeze, as it stirs the leaves with characteristic rustle, is laden with fresh supplies. The breezes which sweep the crowded streets and close alleys of our large cities and towns carry off the carbonic acid, away perhaps to the primeval forests of the great Amazons valley, or to the pine-clad slopes of the Alps. Our own grassy hillsides and meadows partake of the supply. Everywhere the oxygen of the carbonic acid is returned to the atmosphere again, to invigorate and regulate the bodies of animals and men. Nearly twenty-five millions of square miles of leaf surfaces are engaged on the dry-land surface of our globe in removing the carbonic acid daily thrown into the atmosphere!

The "raw material" of plants—that made up of the salts, etc., extracted by the roots from the soil, and by the leaves from the air—is *Protoplasm.* It is a living, usually a jelly-like, substance, composed of carbon, oxygen, hydrogen, nitrogen, sulphur, and phosphorus. These form a sort of committee, with power to add to their number. Consequently, when any other element is required by protoplasm—such as soda, potash, lime, magnesia, etc.—it is at once enlisted. This protoplasm is to the cells of a plant what the prepared clay of the brickmaker is to the bricks formed of it. In the construction of the *walls* of cells, however, considerable economy is exercised—only carbon, oxygen, and hydrogen being employed. Hence the composition and identity in the structure of the cell-walls of all kinds of plants contrasts singularly with the variety of the chemical contents of the cells. The *albuminoid* substances of plants, although they assume a good many shapes, are chemically very much alike, their *nitrogen* being to animals, perhaps, their most important constituent. These are usually most abundantly present in the *growing* parts of plants.

Thus far I have been concerned with sketching the *vegetative* life of plants, and the structures which carry on the work. Now let us turn to the not less important structures and functions of *reproduction*. The former are best known to us in the shapes of flowers and fruits.

The chief, nay, the only necessary organs of a flower, are the pistil and stamens. All the rest of floral organs are simply auxiliaries, called in to assist when necessary, but not unfrequently dispensed with altogether, as in the flowers of the Nettle, and many others, all of which are unattractive. We may take the flower of the common Vine as an example, because this flower has few other floral organs, and is very simple in its character. The reader will see the squat bottle-shaped *pistil* occupying the central position, and five *stamens* grouped around it. The ends of these stamens are swollen into bags in which *pollen* is formed. At their bases are glands where saccharine matter or honey is secreted, to tempt insects to visit the Vine flowers. The flowers

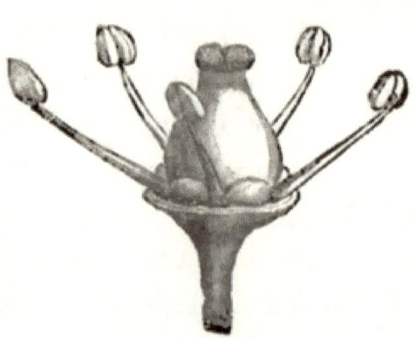

FIG. 11.—A simple flower of the Vine, showing the five stamens surrounding the pistil.

(like many others) also emit a sweet-smelling perfume which convey the knowledge of the whereabouts of the Vine flowers to any insect that may happen to be on the wing.

The number of pistils and stamens varies very much in different flowers, but the pistils are always central, and usually green, so that they can be easily recognised. Formerly these organs were regarded as male and female, and the Linnean system of classification was based on their number, position, etc.

The true flowering plants are now arranged in two groups. Some have beautifully-coloured or sweetly-perfumed flowers, like the Rose, Lily, Sweet-Pea, etc.; others have flowers not conspicuous either for colour or perfumes, like those of the Nettles, Grasses, etc. These two groups are relatively called *Entomophilous* and *Anemophilous* flowers. No more brilliant discovery was ever made in botany than that which furnished the key to unlock the secrets of this floral difference, which is based upon the known fact that "crossing" is not only beneficial, but in most cases absolutely necessary. There are only two agents which are equal to the task of effecting such cross-fertilisation, the *insects* and the wind. The former visit flowers, many of them—such as bees, butterflies, moths, etc.—getting their living in no other manner. Their bodies are covered with hair, and the *pollen*-grains discharged from the ripe stamens stick to them, and are thus carried from one flower to another. As an insect on the wing usually prefers visiting the same kind of flowers (at least bees do), it follows that *crossing* must in this way be brought about. In order to render the services of insects more effective, the pollen-grains of flowers they are in the habit of visiting are usually roughened all over, so as to enable them to stick the better. We can therefore understand the utility of colour and perfume in flowers, and perceive how serviceable to them such qualities must be, inasmuch as they attract insects to fertilise them.

For the same reason we apprehend why grasses

and many other flowers do not possess petals—they are fertilised by the *wind*. Every farmer knows the necessity for dry, fairly windy weather, when "the corn is on the bloom," for then the pollen is carried about. Every acre of wheat produces about 50 lbs. weight of pollen. If it happens to be wet weather, a good deal of the pollen is washed to the ground, and never reaches the *pistils* at all, so that they cannot be fertilised. Unless they are fertilised no seed-corn is produced. The weight or production of the grain crop depends on the number of the flowers in the ears which are effectively fertilised.

When a pollen-grain from one flower has been conveyed to the upper surface of the pistil of another flower, it is held there by a sticky matter which exudes. The pollen-grain begins to sprout, just as if it were a seed, and buds forth a tube which has the power of making its way right to the base of the pistil.

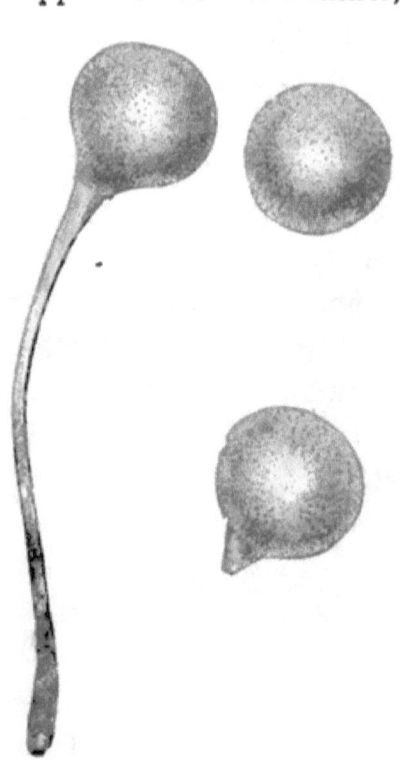

FIG. 12.—Pollen-grains (highly magnified), showing first the pollen-grain, then the latter sprouting and at length budding forth a pollen-tube.

That part is called the *ovary*, and if we examine an unfertilised pistil we shall see its base packed with little whitish-green, egg-like objects called *ovules*. These ovules would never come to anything,

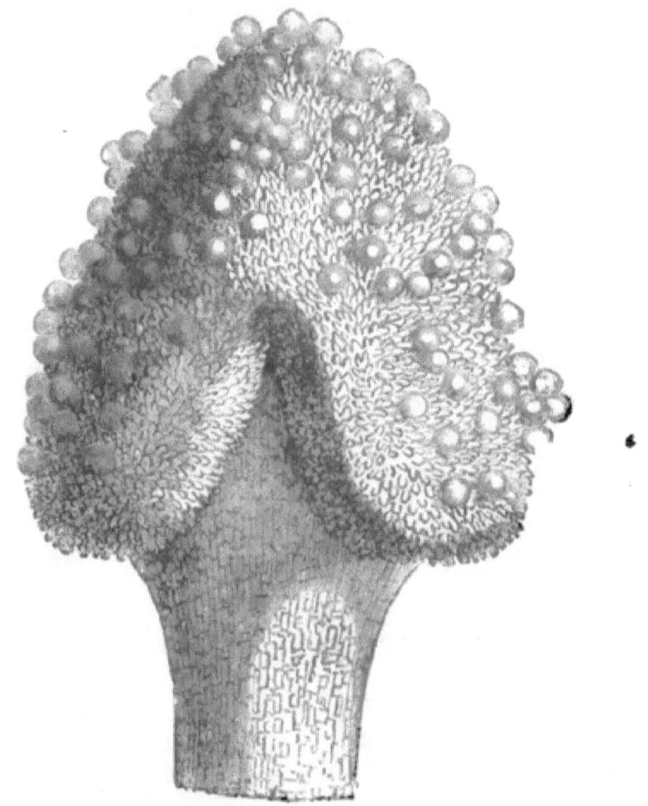

FIG. 13.—Magnified summit of the pistil of a flower, covered with pollen-grains.

and would be thoroughly useless, unless the pollen-tubes made their way right through the tissue of the pistil to where the *ovules* lie. Having reached and penetrated each ovule by a little opening called the *micropyle*, the contents of the pollen-grain are poured out, and the ovules are now said to be fertilised. Each ovule begins to increase in size— in other words, to *grow*. The top of the pistil

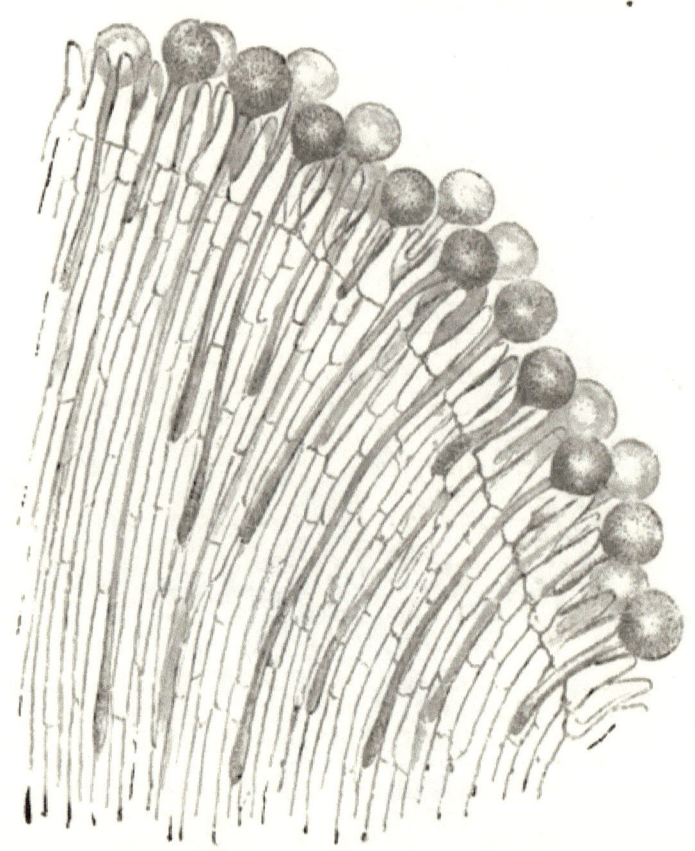

Fig. 14.—Highly-magnified section of the pistil of a flower, showing the pollen-grains shooting forth tubes which penetrate the tissue.

usually withers, and the stamens always; and the vital energy of the plant, or what is left after expending so much in insuring cross-fertilisation, is thrown into the act of developing the ovules into *seeds*.

CHAPTER IV.

WOOD-CRAFT.

THE members of the vegetable kingdom exist amidst a mutual competition compared with which the worry of human life is a state of rest. This battle for existence goes on in addition to the attacks of numerous enemies of all kinds, which prey upon plants. Consequently, we cannot be surprised to find plants adopting devices to gain a livelihood as well as to escape enemies.

Such an adaptation on the part of plants to their environment has been going on through all those periods of geological time that have passed away since vegetable life first appeared on the earth. The modifications and specialisations of terrestrial vegetation have been increasing in complexity and completeness, and the vast variety met with in existing species of plants are the results of this long-continued and unbroken series of adaptations. Plants have now attained to a higher degree of mechanical specialisation than they ever before experienced.

I have chosen the heading of the present chapter to express certain special devices adopted by the higher plants to grow, escape enemies, and propagate their kind.

The giants of the forest have won in the battle of life by sheer strength and bulk. Their huge trunks lift the branches on high, and enable the leaves to hang out like green banners in the sunshine and the breeze. There is even a competition among the trees, individuals as well as species, which shall grow tallest. Who that has wandered through an unkept British forest has not been struck by the keen struggle going on among the trees for place and height? Not even in our overpopulated cities and towns are we so smitten with the fact that the weakest goes to the wall! One cannot explore a primitive wood without feeling that plants are as selfish and greedy as animals, not even excepting man!

The dense shade, which usually prevails in old woods and forests, allows but a scanty existence to those kinds of plants which have not been fortunate enough to accumulate sufficient woody tissue, to build up their stems into tree-trunks. The carbonic acid of the adjacent atmosphere is seized upon, firsthand, by the outstretched stratum of foliage overhead; and even if there were more of it, the umbrageous canopy prevents that admission of sunlight which, as we have already seen, is necessary to the stimulating action of the *chlorophyll* in leaves. Hence the leaves of the herbaceous plants in our woods and forests are often much divided, as in many species of Umbelliferæ, Ferns, etc.; or else plants growing under such circumstances have to be humble-minded, put forth few leaves, and be content with a little. In the vegetable kingdom, therefore, Lazarus can only exist by the crumbs which fall from the rich man's table!

Grim vengeance, however, is often taken. Underneath every Oak, Beech, and Elm, where the abounding foliage of the individual tree prevents herbaceous growth, the progeny of the same tree are sprouting, trying to grow, hustling and jostling each other, dying year after year; for only one plant can eventually take the place of the parent-tree, even after waiting perhaps half a thousand of years!

The Pines—oldest, doubtless, of all exogenous woody trees—are more tyrannous still. Their needle-shaped leaves, abounding in silica, fall to the ground, and mat it so that scarcely a Lichen or Moss can find sufficient foothold for its abiding-place. No other plants so effectively carry out this dog-in-the-manger policy.

Yet in long-cultivated England we practically know comparatively nothing of the greediness of the vegetable kingdom. It is seen in its intensest form in tropical regions, where the stimulating action of the solar light and heat (assisted by an abundant moisture) is at its maximum. In equatorial forests, especially those of ancient standing, the bush-ropes strangle, the parasites bleed, and the epiphytes hang on for a living. Travellers have to hew their way through the dense, dark, tough, selfish mass of vegetation, just as miners have to force a passage through the living rock; but armchair travellers cannot realise sufficiently vividly the actual state of things. Still, even in our own mother-country, we are not without numerous members of the British flora which live by the same devices as their tropical brethren. Not a few of them have had their habits ingeniously directed to serve our own purposes; but there are few people who train the Honeysuckle, Ivy, Virginia Creeper, and Clematis to grow on the walls of their houses, who are aware of the numerous biological and physiographical circumstances, extending and accumulating during geological ages, which have enabled these plants to serve such æsthetic ends!

To the sympathetic botanist there is something touching in the manner with which certain plants accept their fate. The gorgeous-flowered Rhododendrons of the Himalayas have been trained by long æons to be able to exist under umbrageous arboreal foliage, and we have introduced them into our shrubberies because of this (to us) "valuable habit." Most of the shrubs and plants we utilise in our gardens on account of their growing in the shade, have learned the lessons of their specific lives in the same school of experience. No genuine

FIG. 15.—Butcher's Broom (*Ruscus aculeatus*), with flattened branches or *cladodes* doing duty for true leaves.

botanist can regard that remarkable and unique British plant, the Butcher's Broom (*Ruscus aculeatus*), with other than intense interest. It is the only species of *woody* monocotyledonous plant we have in England—the only representative of the woody-stemmed Palms, etc., of the tropics—submissively growing beneath the shade of trees which came into existence ages after its own family had occupied the proud position of aristocrats in the vegetable world. What a story of quiet suffering and struggling with these plutocratic newcomers does the fact that the Butcher's Broom has no leaves, but only *cladodes*, tell us! Leaves with it have long since disappeared. Profitable as they usually are, the plant could not make ends meet; and so the branches flattened themselves, became covered with *stomata* (or carbon-feeding mouths), and performed, and do still perform, all the functions of true leaves. Edwin Waugh, the well-known Lancashire poet, expresses a great botanical truth, although in the broadest vernacular, in his lines—

" For Daisies liven weel
Wheer' Tulips connot grow."

Of those lower members of the vegetable world which, like Uriah Heep, prefer to be humble—such as Mosses and Lichens—we cannot say much. They never appear to have "had their day," like other great orders of plants. The existing Club-mosses once grew to the magnitude of forest trees, as *Lepidodendron, Sigillaria,* etc.; even the diminutive

Horsetails of our waste grounds, damp woods, and ditch-sides, in Carboniferous times acquired both woody growth and a height that would overtop most modern British trees. Grasses still grow to tree height in tropical regions, as witness the Bamboos; and the Ferns, under such favourable circumstances as New Zealand and other places afford, have managed to retain their arboreal supremacy, although elsewhere they have had to submit to fate and descend to the level of humble plants—just as we find many "Howards," "Talbots," "Goodwins," etc., now working among our ordinary population for less than a pound a week! Humboldt thought that in primeval times Lichens might have had arboreal dimensions and magnitude, and that geologists would ultimately find them. But this prophecy of nearly half a century ago has not been realised. Perhaps the Lichens are taking it out in *time* instead of size, for no plants, not even the Californian "big trees," extend through a longer existence than these "gray patches" on the rocks of our hills and mountains. As to the Mosses, nobody has yet expected they would find primeval specimens 50 feet high—small in comparison with Palæozoic Horsetails. The extraordinary geographical distribution of even individual species of Mosses may be regarded as quite sufficient, and that mere individual bigness is not worth striving for in the face of such a fact.

Individual growth is of the first importance; for

if this fails there can of course be no reproduction. No growth can take place without sunlight, for growth is but the transformation of solar energy. Where all are striving to attain the same end, there must be a tailing series from the first to the last. All plants cannot acquire the magnitude of forest trees. That would be a success limited to comparatively few; but such conquerors would do their best to retain the gained position from one generation to another. The Tree-ferns have preserved their arboreal habit since the Devonian period, as have also the Pines. Club-mosses are dying out; they have lost all claim to arboreal magnitude (unless we except the erect New Zealand species of a few feet in height). Indeed, we may regard them as trading on the fortune of their distant geological ancestors!

But what sheer strength and robustness cannot effect, vegetable wood-craft may. The reason why trees grow to great heights is that their leaves may reach the sunlight. A great store of tissue has to be laid by for this purpose in the tree-trunk, and every year witnesses an addition to it, so that it may sustain the burden of the leaves and branches. Hosts of plants, however, have found it possible to get to the light without being at the expense of building up these huge stems. Nevertheless, they manage to overtop the highest trees even in tropical forests; and this they do with wonderfully little expenditure of their organised material. They even succeed in throwing their own foliage outside that of the trees and shrubs of our forests and green lanes, and thus not only get sunlight and energy first hand, but are also the first to be served with carbonic acid from the atmosphere.

When we remember that such plants make use of the huge forest-trees to achieve these important individual ends, and that they would not have attained the positions we find them in unless high woody trees had acquired their present magnitude first, our admiration for the methods adopted by them increases. The mere statement of this fact carries with it to the mind of the botanist the knowledge that such successful climbing habits have been *acquired.* The devices thus developed undoubtedly partake of the character we should call *sagacious* if animals had displayed them. What shall we term them when they are possessed by plants?

Look at our English green lanes, for instance. Notwithstanding the sturdy hawthorn hedges are so durable in trunk and branches (which latter, by their close compressment, and armed with sharpest of thorns, — whence "quick"- thorn, — seem well able to take care of themselves), — in spite of the dense closeness of such shrubs and the little light allowed to fall within, they are beaten in the competition for vegetable existence. Brambles of several species grow in and out by means of their flexible stems, and eventually fling themselves outside, sheeting the hedges with their pretty leaves. That geologically oldest, perhaps, and most cosmopolitan of living ferns, the Common Bracken (*Pteris aquilina*), interweaves itself; the Goose-grass or "Cleavers" (*Galium aparine*) (well named so by the common people), manages to pull itself up by means presently to be described; Honeysuckle, Clematis, Tufted Vetches, Convolvulus, Black Bryony, White Bryony, Ivy, etc., utilise and overpower the poor Hawthorn, and beat its robustness by their wood-craft!

In tropical and equatorial forests this game is played even more vigorously. The Bush-ropes and other climbing plants wrap round the largest of tree-trunks, twist themselves in and about the arboreal foliage, and eventually reach the highest points, reserving their vegetative power until then, and putting it forth under the most favourable circumstances as to light and heat. In the Central American forests, among such successful schemers may be mentioned *Marcgravia umbellata*, which flattens and moulds its own stem around the trunks of robuster forest-trees, puts forth root-claspers to embrace them, and so raises itself, like a *parvenu*, above those which help it. And eventually, when it has reached the light above, and overtopped the foliage of the tree it climbed by, it throws out branches with ordinary round stems and leaves like other plants—just as if it had not cheated, and strangled, and done all kinds of vegetable crimes before it satisfied its own ends!

I have said that our thickest European tangles cannot afford us the slightest idea of the similar struggle for existence going on in tropical forests. At home the battle is fought out only a few feet above the ground—there it is going on high overhead. Below, the creeping plants are gripping for the tough wrestling match, and as we see them twisting and writhing in and out, as Kingsley says, we cannot but feel they have beaten the woody trees in the battle, and that by sheer cunning. Some tropical genera, like the *Bauhinias* of Brazil among the Leguminosæ, seem to have laid themselves out purposely for climbing; and the faculty is now instinctive, as doubtless would be the habits of athletes if a thousand generations of them were confined to such a mode of life. Wallace mentions one of the most extraordinary of the *Bauhinias* he saw in the forests of the Amazons valley, which had a broad *flattened* stem, that twisted in and out in the most singular manner, mounted to the tops of the tallest forest-trees, and thence hung down in gigantic festoons many hundreds of feet in length.

These climbing plants cannot live below—they must join the great throng of floral life up above. So that, as Wallace also remarks, walking in a tropical forest is like sauntering amid the columns of an empty cathedral, whilst the service is being celebrated aloft on the blazing roof!

Our own "Ivy green" is less cruel and selfish, although we frequently see it growing to such a luxuriance as to throw its own foliage over that of the ancient oak, whose trunk it had utilised in order to get above the ground, and attain to all the advantages it would have possessed had its own stem grown bulky and erect. This height it reaches by special structural organs. Indeed, all the weak-stemmed and weak-trunked plants put forth similar special efforts, which seem to partake more of cleverness than of strength. The Ivy, however, is honest. Its clasping roots do not rob the tree it climbs by of any nourishment or sap—notwithstanding prejudice to the contrary. We doubt whether the Ivy is not often a benefactor—shielding and protecting stems and branches from severe winter cold, by its stratum of non-conducting foliage. The Ivy is not the sole possessor of this particular method of climbing. A species of Fig (*Ficus repens*) in the East Indies so far departs from the habits of its fellows as to raise its weak stems on walls and rocks by means of creeping roots like those of the Ivy. The latter species of Fig has improved upon the Ivy's method, for its roots emit a viscid fluid which acts as an adhesive cement, and assists the plant to hold on by. There is also the sweetly-perfumed *Hoya carnosa*, one of the Asclepiads, common in most conservatories, which develops ivy-like rootlets for attachment, although its weak stem has the power of twining as well. A well-known North American plant, the Trumpet-flower (*Tecoma radicans*), is another species employing adventitious roots to assist it in climbing.

Whilst numerous species of weak-stemmed plants thus raise themselves to a greater height, and put themselves into a better position, by adopting one particular kind of device, we find others which have succeeded in acquiring several kinds. One cannot but be struck with the fact that these mechanical contrivances are not peculiar to any order of plants. They are shared by members of all alike. Many of the species adopting them are mere vegetable adventurers, others are species which "have come down in the world." The flexible stems of the Brambles have been already mentioned as admirably adapted to penetrate the thickets of hedges and shrubs. In doing so they are assisted by their recurved hooks, which act as grappling-irons. But perhaps the most perfectly developed of grappling-iron machinery is that possessed by one of the commonest of our English wild plants, the Goose-grass (*Galium aparine*). Its other name of "Cleavers" well expresses its power to stick on and cleave to anything it rubs against, or is thrown at. A pocket magnifier makes the reason plain. Stems, leaf-stalks, leaves, and fruits — all are crowded with myriads of hooks. No wonder it grows so luxuriantly along our hedgerows, overtopping the hedge itself, notwithstanding its unusually weak stem! A well-known genus of exotic plants is the *Smilax*, whose name is derived from a word signifying a scraper, in allusion to its stem being roughed with curved prickles, by means of which it climbs, grappling-iron fashion. Some species have managed to develop a more highly organised and sensitive vegetable contrivance as well, namely, the use of *tendrils*. Various kinds of weak-stemmed Roses possibly use their thorns to climb by, and by which to keep their places among other plants. The Wild Hop, in addition to its more natural and successful habit of *twining* round other stronger plants, is not without hooks, which appear to be a more recently acquired mechanism. Several species of a genus of plants notable for their climbing habits, the *Cobœas*, have horny hooks, and it is doubtful whether any group of plants have become so highly specialised for climbing as they. Only give them a rough wall-side, and the tendrils will catch fast hold of the crevices, and support the branches without any assistance, except those of the more recently developed hooks.

Twining appears to be the simplest and most primitive method of climbing among plants. It is due to the weak and rapidly-developing stem growing a trifle faster on one side than the other—just as a carpenter produces any degree of curl in his shavings according as he presses his plane a little more on one side than the other. This twining habit, however, does not manifest itself until the plant has grown for some time. At first the stem is erect, and puts forth its leaves at the joints, or nodes, in the ordinary way; but when it has developed two or three such joints all its natural erectness gradually leaves it. The spaces between the next joints grow faster, and more unequally, so as to cause the stem to curl and revolve. Of course the amount and rapidity of this curve will be proportionate to the unequal growth. The plant appears as if it were feeling for something to twine round, and when it has found such an object it at once begins to clasp it. The habit of growing erect at first, even among such confirmed twining plants as the Honeysuckle, Tropæolum, and Hop, etc., plainly shows us that twining has been gradually acquired. All sorts of plants alike have adopted the practice, even Ferns and Adder's-tongues; for it is highly developed in the well-known *Lygodium*, and is also indulged in by *Ophioglossum japonicum*, which latter climbs by means of its twining leaf-stalks. The habit is indulged in in varying degrees. Some plants only occasionally require it, such as the Bitter-sweet

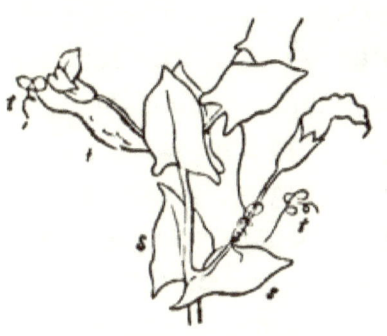

FIG. 16.—Yellow Vetch (*Lathyrus aphaca*), showing leaf-like stipules (*s*), and true leaves converted into tendrils (*t*).

(*Solanum dulcamara*), some Asclepiads, etc.; these appear to be just adopting the plan. Others climb only in the summer time, like our climbing Buckwheat (*Polygonum convolvulus*). Some species twine one way and some another—with the sun and against it, according to the popular method of defining it. A few, as *Hibbertia*, twine both ways, and so, as Darwin says, this plant is admirably adapted for rambling in and out through the Australian scrub.

It is interesting to notice how, when the stem of a plant does not twine, or only acquires the disposition and power to do so after sprouting the first two or three joints, the branches will commence the habit much later on. Only the lateral branches twine in *Tamus elephantipes*, and in *Periploca Græca* the habit is confined to the uppermost shoots. A large number of species climb because the foot-stalks of their leaves are able to twine. This is particularly noticeable in that greenest and loveliest of the climbing plants which, in the summer months, festoon our green lanes, the Traveller's-joy (*Clematis vitalba*), whose bare and withered stems, even in winter, may still be seen retaining the hold they gained months before. A rarer and much weaker plant, only to be met with in thickets, is the climbing Fumitory (*Corydalis claviculata*), which uses both tendrils and leaves to climb by. Its near relatives, the true Fumitories, are backward in developing their climbing powers, although one species, *Fumaria capreolata*, is fairly on the way. Among other well-known leaf-stalk climbers are species of Bignonia, Solanum, Maurandia, Lophospermum, Tropæolum, etc.

The highest degree of specialisation for climbing purposes, however, is the development of *tendrils*. These may be, and usually are, modifications of stipules, leaves, and branches. Be that as it may, tendrils are organs differentiated to perform a special kind of work for the benefit of the plants possessing them. They may be found in nearly every stage of development, from single to even branched tendrils. All of them, however, are distinguished by their *sensitiveness*, which reaches its climax, perhaps, in *Passiflora gracilis*. Some of this class of climbing plants have branched tendrils possessing suckers at their buds, like the well-known Virginian Creeper (*Ampelopsis hederacea*). The suckers exude a natural cement, which, however, does not form until the sensitive tendril-tips have first found a good foothold. Similar discs for the tendrils to hold on by are developed by *Bignonia capreolata*, and a species of the Cucumber family. Some plants have tendrils provided with hooks at their tips, like the *Cobæas*, which they insert in chinks and crevices. All tendril-bearing plants are further remarkable for the help afforded these sensitive organs by the stems. Instead of being rigid, as we have seen is the case in the first two or three joints of twining plants, the young internodes adopt the opposite plan, and revolve slowly, so as to bring the tendrils into contact with some object around which they can clasp. Instinct could not go further, even in a highly organised animal.

Of course these vegetative devices are variously successful, just as the facilities for performing them vary. But in all instances an end beneficial to

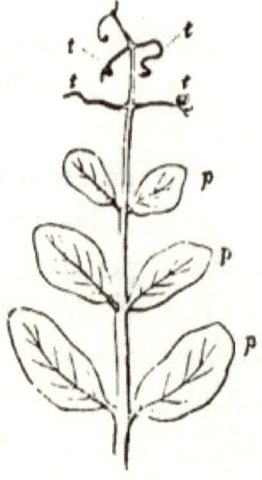

FIG. 17.—Leaf of Pea, showing gradual decrease in size of leaflets (*p*) until the tendrils (*t*) are formed in their places.

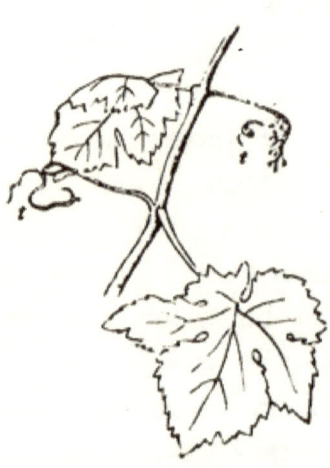

FIG. 18.—Tendrils of a Vine (*t*) replacing leaves.

the plant employing them is gained ; and in most the climbing plants have beaten the tall, strong, woody trees and shrubs in the contest for light, heat, and gaseous food. They have conquered by sheer wit, or the equivalent of it ; and there is little doubt if animals had been similarly successful in their endeavours to achieve a certain end, we should have spoken of them as " clever."

CHAPTER V.

FLORAL DIPLOMACY.

IT is the reproductive organs of plants which usually afford most pleasure to our æsthetic sense. The details of these parts have already been referred to. Most species of plants could not be perpetuated without the act of flowering, although a comparatively few have discovered a knack of developing new individuals without resorting to this generally adopted plan. Perhaps it would be as well to note the exceptions first.

The Potato suggests itself immediately to one's mind as a type. It is an underground bud, richly supplied by the parent plant with starch and food, just as other plants supply the seed-lobes (*cotyledons*) with food-stuffs. Its "eyes" are the young sprouts, and we can actually separate a single potato into several individuals when planting, if we take care not to cut across any of these parts, but between them. All tubers have the same privilege as the potato, of being able to reproduce the species by other means than seeds. Bulbs, corms, etc., are not included in this class, for they are really arrangements for continuing the life of the individual, rather than of perpetuating the species. We have also "adventitious buds," such as may be seen on the crown of pollarded trees, or breaking out from the bark of the Elm and Maple. Not unfrequently the roots may form them, as in *Pawlonia, Maclura aurantiaca*, and the better known *Pyrus japonica*. Indeed, under favourable circumstances this class of buds, each of which develops into an individual plant, may be produced from any part of the vegetable organism.

In the "Air-plant," so called by cottagers in whose windows we see it suspended (*Bryophyllum proliferum*), the thread-like roots hang down, with young individuals sprouting from the nodes. These, as they drop, commence their individual lives without having had to undergo the ordinary process required of embryos produced through the agency of seeds. Such a habit may be regarded as the equivalent of "fission" among the lower animals. Indeed, among many phanerogams or true flower-bearers, the leaf-buds may be separated from the parent plant, and reared into new and distinct individuals. As a rule, "adventitious buds" are the result of unusual vegetative activity—the plant has accumulated more energy than it can expend by the ordinary process of flowering. As an illustration of this I may mention "proliferous" Ferns, which bear young specimens on the upper surfaces of their fronds. They are not at all unfrequent in well-kept ferneries, where there is a superabundance of vegetable food.

The Strawberry affords us an instance of propagation by means of peculiar creeping roots. Everybody is acquainted with the shoots thrown out by its roots in gardens. Strawberries grow more luxuriantly there than on the wild hedge-banks, for more nutriment is present; but the habit was originally acquired in the desperate battle for vegetable life which takes place on every bank exposed to the sun. At each joint of these Strawberry "runners," as they are called, small leaves sprout, and the gardener can propagate a new plant from them. "Couch-grass" and many other plants belonging to widely separated orders have found out a similar plan. Adams and Leverrier discovered the planet *Neptune* almost simultaneously, each without the knowledge of the other;—plants in their life-histories have not unfrequently hit upon the same lucky devices!

Not unfrequently, on very succulent leaves, similar accidental buds are formed. Some plants—like the little freshwater *Hydra* among animals—have even learned to take advantage of accidents that would destroy other individuals, and to propagate their kind thereby. Mr. Herbert Spencer points out that *Bryophyllum calycinum* has succulent leaves, borne on very brittle leaf-stalks, which are easily snapped by the wind, and otherwise broken. But when the fragments fall to the ground, each sends forth buds which root themselves and grow into independent plants. Another species of the same genus, *Bryophyllum proliferum*, habitually breaks its stem near the base, even when it is in flower; so that it cannot ripen its seeds. To compensate for this, we find young plants budding out of the foot-stalks of the flowers, and growing up into separate plants. The leaves of the common Cuckoo-flower, *Cardamine pratensis*, often form adventitious buds when lying, as they usually do, upon wet ground. Although bulbs are formed more particularly to continue the life of the individual plant, year after year, still they have learned in some instances to do more. Thus in Tulips, Crocuses, Daffodils, and many Lilies, they form succulent scales, which succeed in developing new plants. Many other devices could be mentioned by any intelligent horticulturist, whereby new plants are developed otherwise than through the agency of seeds. The ease with which most plants are propagated by means of *shoots*, stuck in the ground at the proper time of the year, best illustrates this secondary method of reproduction.

But, after we have made full allowances for these interesting bypaths, it still remains that inflorescence is the universal plan by which flowering plants are propagated. Flowers, fruits, and seeds are indissolubly associated with the perpetuation of every kind of highly-organised plant. The necessary structures of a flower, stamens and pistil, have already been described (Chap. III.), and we there saw that the rest of floral organs are auxiliaries, although to people unacquainted with botanical details a flower has no charm without brightly-coloured and attractive petals; and horticulturists do their best to convert the true floral organs, stamens and pistils alike, into petals—whence the origin of the double flowers seen in all gardens.

Mr. Herbert Spencer long ago pointed out that the act of flowering must of necessity take place where the energy conducive to growth is balanced by the forces which resist growth. At this balanced point germs can be produced and easily thrown off, because nutrition to the individual plant must there be failing. Hence flowers are generally borne at the terminal ends or shoots of branches, where nourishment is least, and not most abundantly supplied. It may seem to many people very strange to regard flowers as lower in organisation than leaves; but we have seen such is the case from a structural point of view. Floral organs are in reality *aborted* leaves, although, since flowers came into existence, and have had to adapt themselves to their organic and inorganic surroundings, they have done the best they could with them, and even converted and modified their auxiliary and other floral organs, until they are now the chief marvels of vegetable organisation, and the delight of every cultivated mind.

To thoroughly understand the amazing variety—in size, shape, colour, and perfume—among flowers we must consider the whole question from their own point of view. Hitherto we have taken it for granted that flowers were created chiefly, if not wholly, for human delight or use, and have, with an unintentional irreligion due to ignorance, expressed our wonder why all plants did not administer either to our utilitarian or æsthetic needs! It is not necessary here to mention the ingenious guesses which have been put forth to account for the apparent anomaly of unattractive, useless, and even poisonous plants; for no botanist now doubts that such speculations are all wrong. If modern botany had done nothing besides abolishing these crude views, it would have a claim to gratitude; but it has done more—it has taught us to regard plants as *fellow creatures*, regulated by the same laws of life as those affecting human beings themselves!

To understand the structure of flowers, therefore, we must first of all consider how far they are useful to the end for which flowers were developed. This, of course, is the production of seeds, each of which contains an embryonic individual plant.

The process by which an ovule is converted and developed into a seed has already been sketched.

Such process is called "Fertilisation." No fact in modern botany has been better proved than that "crossing" is beneficial—in other words, that the seeds of flowers whose pistils have been fertilised by the pollen brought from the stamens of another flower, and especially from that of another plant, produce better and stronger individuals than the seeds would have ripened into if the pistils had been fertilised by pollen from the stamens of the same flower.

The strongest efforts of the floral world are put forth with the view to flowers being "crossed." In every country, in every geological period since flowers were first differentiated, the competition has been going on. It is now in the midst of its fiercest and intensest action, for never before, in the entire history of our planet, were the agencies involved in it so complex. Here the race is to the swift, and the battle to the strong. Those best handicapped in the race win—the hindmost linger on as best they can. The botanist recognises self-fertilisation or only occasional crossing, in the weakly, dwarfed, small-flowered species, most of which are best known by the popular name of "weeds," a term expressive of their uselessness and usual lack of floral beauty and strength.

Between such individuals and the most complex of floral mechanisms we find every possible stage of organisation. It is because of this gradation that flowers have such marvellous diversity. The desirability of being "crossed," because of the greater certainty that more and stronger seeds will be the result,—from which may possibly spring a succeeding generation of plants even more robust than their ancestors,—is as important an element in the life of an individual plant as it is for man to work and save, so that his children may be better placed in society.

But if all plants adopted the same contrivances to be crossed, there would be a painful floral monotony. Some flowers might then be large and others small, but there would be nothing of the amazing varieties of floral shapes, colours, and perfumes, we see everywhere around us.

No person possessed of even the most rudimentary knowledge of flowers can help grouping them under two kinds—attractive and non-attractive. It has just been mentioned that many of the latter are so because they are self-fertilised. But this latter class is slenderly represented, so that we may leave it out of our calculation just now. Observe the number of plants, shrubs, and trees, which bear flowers so thoroughly deprived of all the organs and qualities with which we associate them, that many people express surprise when they hear they bear flowers at all. Among our British plants such species as the Dog's Mercury, the Nettles, Wild Hop, Pellitory, Hazel, Birch, Poplar, Alder, Oak, etc., as well as an immense number of species of grasses, sedges, and rushes, roughly illustrate what a large number

Fig. 19.—Annual Dog's Mercury (*Mecurialis annua*).

bear unattractive and unappreciated flowers.

Fig. 20.—Pellitory of the Wall (*Parietaria officinalis*). A. Pistil with tufted stigma.

Then follow the mighty host of what most people

would be inclined to call *real* flowers, varying from the smallest sized, feeblest coloured, and unperfumed of the species blessed with occasional crossing, to the gorgeous floral aristocrats which cannot exist without it, and must be supplied with it. Between the Chickweeds, and the Peonies and Magnolias, there is intercalated a series so beautifully graduated according to their needs that it would be difficult to find a genuine hiatus, could we only bring all the species of flowers in the world together for comparison!

Still, in the mind of the observer, this last group of flowers would be distinguished from the first mentioned by their possessing *petals*. Popular opinion associates those usually brightly-coloured and always beautiful parts with flowers themselves (as we may see in the artificial kinds used for feminine adornment), and it cannot conceive those are real flowers which do not possess them.

In reality, as we have already seen, a genuine flower requires only *stamens* and *pistils*, and these are present in varying numbers, as the classification of Linnæus shows us. They are the only genuine reproductive organs of flowers,—all the rest are mere "helpmeets,"—and yet these auxiliary parts of flowers have always affected mankind most.

Is there no explanation to account for the two great divisions of unattractive flowers, and flowers rendered attractive by auxiliary aids and helps?

So important is it for flowers to be crossed that Darwin revived and somewhat modified an old saying, by declaring that "Nature abhorred self-fertilisation." In our country, at least, if flowers are to be effectually crossed, only two agents can perform such a function—the wind and insects. The former knows nothing of beauty of colour or sweetness of perfume. Flowers habitually crossed by the wind would literally "waste their sweetness on the desert air," if they were to expend their much-required energies in developing such attractions. Hence all wind-fertilised or *anemophilous* flowers are unattractive. On the other hand, insects are possessed of such a marvellous sense of smell that it is probably their chief sense, and that by which they gain their knowledge of the world around them. It is to this acute sense the perfumes of flowers appeal, whilst the mass of colour displayed by the petals assists in directing the insects to the flowers. Meantime the shapes of the latter are generally related to the ease with which insects can alight upon them. Botanists have done well in grouping all the flowers of this class under the term *entomophilous*, or "insect-lovers."

Still, there is no distinct hard-and-fast line between these two divisions, extreme as they are; for we find numbers of species of plants whose flowers resort to both agencies for crossing, as those of the Docks and Plantains, for instance. Some of the insect-fertilised group are gradually giving up, or have already given up, their ancient habit, and are taking to self-fertilisation, like the Bee Orchis. Many of the grasses, rushes, etc., and such peculiar species as the *Pringlea*, have become wind-fertilised ; whilst some wind-fertilised kinds have advanced a stage, and are visited by insects because their flowers have become more attractive, as the catkins of the Willow. It is probable that, in the order of their geological appearance upon the globe, the wind-fertilised (*anemophilous*) flowers preceded the insect-fertilised (*entomophilous*). It is very certain the latter have increased more abundantly than the former within the most recent of geological periods, because that is precisely the time when insects have multiplied and developed. And with the multiplication and differentiation of insects has gone on, *pari passu*, the specialisation of flowers.

No department of modern botany is more delightfully attractive than that devoted to the study of how both of these great divisions of plants have succeeded in employing wind and insect agencies to assist in cross-fertilising their flowers. This end has been accomplished with various degrees of success, just as animals differ among themselves in the amount of sagacity they possess. But the beaver does not display more instinct in constructing its dam, the bee in the manipulation of its geometrically-shaped comb, or the white ants in the erection of their habitations, than flowers do to attract insects to themselves, to point out by means of "honey-guides" where the nectar lies, and to detain them there by mechanical contrivances until the pollen is discharged upon their hairy bodies, or brushed off them if already there! Some of the most important of these devices will be noted in the following chapter.

CHAPTER VI.

FLORAL DIPLOMACY (*continued*).

LET us commence with a device common both to wind-fertilised and insect-fertilised plants, so well marked that Linnæus founded a class upon it, and thus grouped together the most widely separated species of plants, which had nothing in common except that their ancestors had hit upon the same simple method to prevent self-fertilisation. The contrivance is based upon the principle that "prevention is better than cure!" If the flowers of a plant bear stamens only, and others pistils only, it is evident there can be no danger from self-fertilisation, and the chances of crossing are then rendered almost certain. If this principle is carried to the furthest point, and an arrangement is made that the female (or *pistillate*) flowers shall be borne on one plant, and the male (or *staminate*) flowers upon another, crossing is absolutely insured. Should such flowers fail to be crossed, no fruit is borne, and the flowers are then what gardeners call "blind," as in the case of Cucumbers, etc. The process of differentiation which has produced such a result is not yet completed in some species, as the Elm, Horse-chestnut, and Common Ash (*Fraxinus excelsior*). On the last we find ordinary flowers, possessing stamens and pistil, and pistillate and staminate flowers as well. The Common Maple (*Acer campestre*) is even less advanced than the Ash, but it is being specialised in the same direction.

Among our British plants we have numerous proofs that these remarkable adaptations to special and important ends have been slowly and gradually acquired. We infer that the *hermaphrodite* condition of flowers is the natural one; that is to say, all flowers were originally possessed of both stamens and pistils. To prevent self-fertilisation some kinds of flowers bear stamens only, and others pistils only, as above described; and this plan is adopted for a special purpose. On examining such flowers we see no reason to doubt that originally they were like ordinary flowers. The modification is going on in our own times, and is probably suggested to the plants by some change in their external condition of life. Thus Darwin has shown that the flowers of the common Strawberry, which in this country are always hermaphroditic, in the United States are becoming sexually separated into staminate and pistillate kinds. The flowers of the Indian Corn (*Zea mays*) are sometimes unisexual and at others bisexual. Some kinds of Palm-trees bear staminate flowers one year, and pistillate flowers the next. That abundant shrub in the hedgerows of the more southerly parts of England, the Spindle-tree (*Euonymus europæus*), produces flowers in which the stamens are frequently, but not always, aborted; those of the Wild Thyme (*Thymus serpyllum*) are often to be found in the same condition. These facts prove that the alteration and adaptation of flowers to their surroundings is still going on. In the sedges the male and female flowers are habitually grouped together, but in separate clusters.

The differentiation of flowers into pistillate and staminate kinds is carried out in the highest degree in those species which bear them on separate individual plants, shrubs, or trees. To them the old Linnæan term *diœcious* is still applied. Evidently this habit is an advance on that we have been considering, where the floral sexes are separate but borne upon the same plant (*monœcious*).

FIG. 21.—Female or pistillate flowers of Oak.

Among the wind-fertilised or *anemophilous* group of flowers, the adaptation is a purely mechanical one. Some bear catkins, like the Hazel, Poplar, etc., where stamens only are clustered together, the female flowers being quite different and also separate. Nothing could be better adapted to wind fertilisation than the pendant position and structure of the male

FIG. 22.—Male or staminate flowers of Oak.

catkins. They are swayed to and fro by the slightest movement of the air, and their pollen is therefore easily detached when ripe. Moreover, this class of trees and shrubs usually produce flowers before they bear leaves, and in our country the flowers appear generally in the early spring months, when the breezes are strong. The absence of leaves is beneficial, for there is no mass of foliage to interfere with the wind carrying the pollen to the female flowers, as there would have been had not these plants selected the best time for inflorescence. Many other wind-fertilised plants flower early in the year, as the Dog's Mercury, etc., although these have no such necessity for limiting themselves to this early season; for, being humble herbaceous plants, they have no great amount of foliage to mask their flowers and hinder their being crossed. The Hops take their time —leafing first, masking the hedgerows they have conquered by wood-craft with their beautiful leaves, and then placing their flowers (always on separate plants) triumphantly in the best spots for the summer breezes to carry the pollen, and to act as their marriage priests. In the grasses, another device is in vogue. The pollen bags, or *anthers* of the stamens, ripen within the closed doors of the flowers, protected from wet and chance cold by the waterproof, chaffy scales. When the pollen within the anthers is thoroughly ripe, the stalks or filaments which bear them grow very rapidly, and so lift or thrust the anthers outside the flowers, where they dangle so that the slightest puff of wind detaches and carries away the pollen.

All genuine wind-fertilised flowers—big and little, herbaceous, shrubby, and arboreal, no matter to what order they may belong—have evolved a similar generalised structure for their pistils. As the latter

Fig. 23.—Barren Brome-grass (*Bromus sterilis*). *a*, Spikelets; *b*, Flowering glume.

have to catch the stray pollen-grains carried by the wind, it is evident that some net-like or feathery shape would be best suited for such a purpose. Some modifications of these patterns we accordingly find—pistils, plumose or feathery, tufted, fimbriated, etc.,—always a specialised organ, admirably adapted to catch straggling pollen-grains. And as these feathery or tufted parts are glutinous, there is little chance of the pollen-grain being blown away again, when once entangled in the living mesh. There it remains and sprouts, and the world gets its "daily bread" by means which at first sight appear purely accidental.

The extremer forms of wind-fertilised and insect-fertilised flowers have been separated so widely, and their differences are so extreme, that even the shapes and structures of their pollen-grains are different; and it is not a difficult matter to tell, from the microscopical examination of a single unknown pollen-grain, whether it has been produced by an inconspicuous or a conspicuously-coloured and perfumed flower. In the wind-fertilised flowers the pollen is generally produced in great abundance, and it is always very light, so that the wind can easily blow it about. Usually their pollen-grains are many-sided, so that more than one surface can be presented to the wind. The pollen-grains yielded by highly-developed insect-fertilised flowers, on the other hand, are usually roundish and oval, and frequently have their surfaces roughened by tubercles, bars, spines, etc., which enable them readily to adhere to the hairy bodies of bees, butterflies, and moths, so that they can be thus easily carried from one flower to another.

A simple device to prevent self-fertilisation is adopted by many flowers. It is evident that if the stamens and pistils of a flower do not ripen at the same time, there is no danger. Accordingly, we find some plants in which the *pistils* ripen first, and these are accordingly distinguished as *proterogynous*. The Figwort, Birthwort, etc., prefer this plan. In the Mallow, Geraniums, Willow-herbs, Gentians, Campanulas, and numerous others, the *stamens* are the first to ripen. Such flowers are called *proterandrous*. Not a few plants adopt one or other of these ingenious

FIG. 24.—Carnation, showing the ripe pistil.

devices, and are monœcious and diœcious as well. In their case self-fertilisation is absolutely impossible, and floral sagacity reaches its highest water-mark.

Certain flowers, such as the Primrose family, manifest such an abhorrence of self-fertilisation that their own pollen appears to act almost like poison

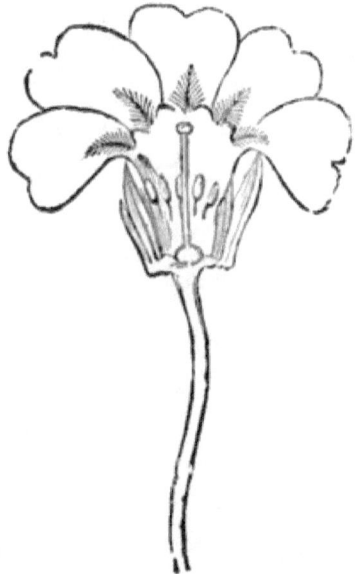
Fig. 25.—Common Primrose.
Pin centre.

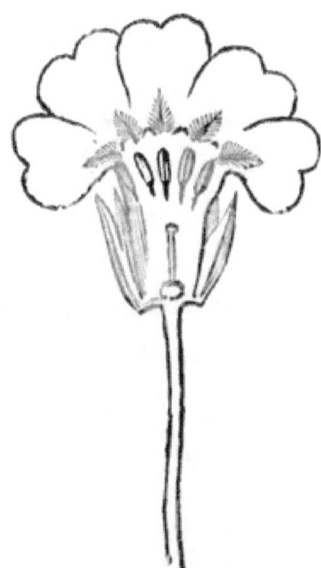
Fig. 26.—Common Primrose.
Rose centre.

to their pistils! Some species prefer to be fertilised by the pollen of another species, and to produce hybrids, rather than be fecundated with pollen produced by the stamens of the same flower.

Hardly any other term than diplomacy so well expresses the means by which certain kinds of flowers have even altered the positions and shapes of the stamens and pistils, so that crossing may ensue. In the Primroses this is seen to perfection, for the flowers of one plant have the stamens situated just at the point where, in those of another plant, the stigmatic surface of the pistil appears. This contrivance has been established, so that an insect may be dusted with the pollen from the Primrose flower, precisely on that part of its body which will come into contact with the pistil in another flower. Common people have long distinguished

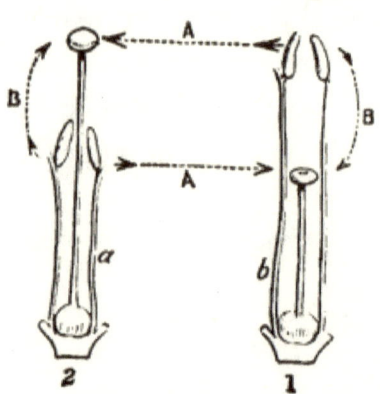

FIG. 27.—Fertilisation of Primrose (*Darwin*), showing how the long stamens fertilise the long pistils, and the short stamens the short pistils.

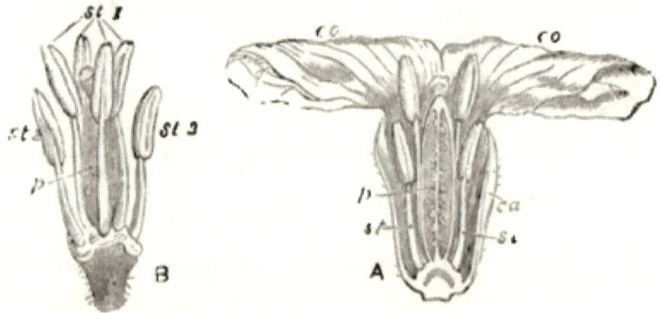

FIG. 28.—Structure of Wallflower. The different-sized stamens (*st.* 1 and 2) may be regarded as an organic change tending to dimorphism.

such Primroses under the names of "pin-eyed" and "Rose-eyed;" but the term *Dimorphism* is now applied to the phenomenon, which is in vogue among

FIG. 29.—Purple Loosestrife (*Lythrum salicaria*).

numbers of our wild flowers, and is therefore well known among botanists.

A smaller number of flowers have proceeded a stage further, and developed *three* lengths and sizes both of stamens and pistils, as in the Purple Loosestrife (*Lythrum salicaria*),—famous for Darwin's lucid examination and lucubration thereon. This additional specialisation of the floral organs for crossing purposes is called *Trimorphism*. Singularly enough, in trimorphic flowers the differentiation is carried to such a degree that the different-sized stamens produce different-sized and coloured pollen-grains.

The great variety in the *colours* possessed by

flowers is now regarded by some botanists as evidence of the degree and status of their development. Dr. Müller and Mr. Allen think that yellow is the original and primitive colour, and that white, red, purple, and blue, were assumed in the order here mentioned. Consequently yellow flowers are nearly always simple in structure, and usually loose-petalled, or *polypetalous;* whereas blue and purple flowers are generally more highly specialised in form, and *gamopetalous*—that is, they have the petals of the corolla united in a single piece, as in the Campanulas and Common Foxglove.

White and light-yellow flowers not unfrequently keep open until late in the evening, or even throughout the night, when their corollas are of course more conspicuous than they would have been if distinguished by any other colour. There is a much larger number of species of night-flying Lepidoptera, or moths, than of the day-flying butterflies, and none of them can exist on any other food than the liquid nectar of flowers. Consequently, the number of light-coloured flowers habitually fertilised early on summer evenings must be enormous. No wonder, therefore, that some species, in every part of the world, have so laid themselves out that they open very little, if at all, in the daytime, but only at night. Many species of such flowers have obtained for themselves botanical names indicative of such a habit, as the Evening Primrose (*Œnothera biennis*), the night-flowering Campion (*Lychnis vespertina*), etc. The odours of these flowers are always emitted most powerfully at night; and *white* flowers bear away the palm in the competition with those of other colours for including the largest number of species possessing sweet perfumes. This of itself is a significant fact, in view of moth-fertilisation.

One of the latest discoveries of the eminent botanist, the late Dr. Hermann Müller, was to explain the cause of *double* colours in flowers, such as are most strikingly shown, perhaps, in our Viper's Bugloss (*Echium vulgare*), but which are also well known in the various species of Forget-me-nots (*Myosotis*), the Lungwort (*Pulmonaria officinalis*), Comfrey, and many other species of the *Boraginaceæ* —which natural order seems to have distinguished itself by its floral development. In the Viper's Bugloss and the Lungwort the flowers are red and blue. The former colour is assumed *first*, and blue as the flowers get older. Dr. Müller proved from examination that all the blue flowers were empty of honey, and that the stigmas of their pistils were supplied with pollen—in fact, they had all been visited and fertilised by insects. He therefore concluded that the blue colours, whilst increasing the conspicuousness of the floral clusters, at the same time indicated to such intelligent bees as *Anthrophora* which flowers they should restrict their visits to, so that neither they nor the flowers might suffer from the loss of time that would ensue if insects were to go blunderingly to every flower and find out for themselves whether it

FIG. 30.—Common Comfrey (*Symphytum officinale*).

had any honey (or had been fertilised) or not! In many flowers belonging to other orders the colours of the petals change as they get old, as, for instance,

84 SAGACITY AND MORALITY OF PLANTS.

in the Hawthorn, from white to a defined and sometimes strong tint of red, and in the Little Celandine (*Ranunculus ficaria*), which bleaches into a white. May not all the faded or fading colours of flowers indicate to intelligent insects which have been already visited, seeing that the usual thing that takes place when the pistil has been fertilised is for the auxiliary parts to fade and die off, for their work is done, and they are no longer wanted?

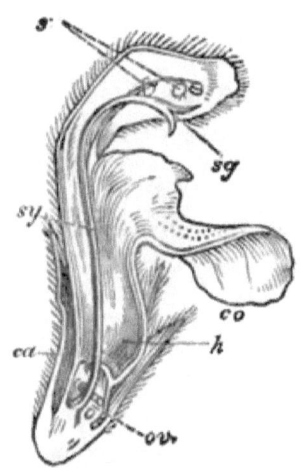

FIG. 31.—Section of Dead-nettle, showing the various parts of the flower, and the labium (*co*) for insects to alight upon.

The mechanical constructions of those all-important floral organs, stamens

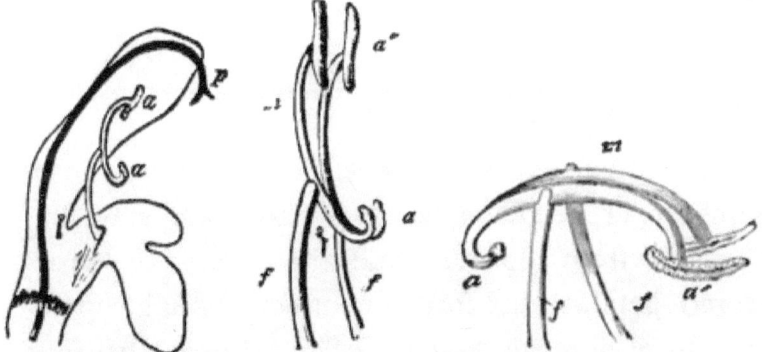

FIG. 32.—Flower and stamens of Salvia. The latter show the movable connective (*m*) by means of which the stamens (*a*) are brought down on an insect's back.

and pistils, in the highest developed of flowers, are worthy of the most careful attention. Let us take such natural orders as the *Labiatæ* and *Scrophulariaceæ* for instance. In no other group, perhaps not even excepting the ingeniously constructed *Orchidaceæ*, do we find such a blended series of high development. First of all, their petals are not only united into one piece (*gamopetalous*), but the different parts have developed unequally, so that the shape of the flower is what botanists term *irregular*. Of

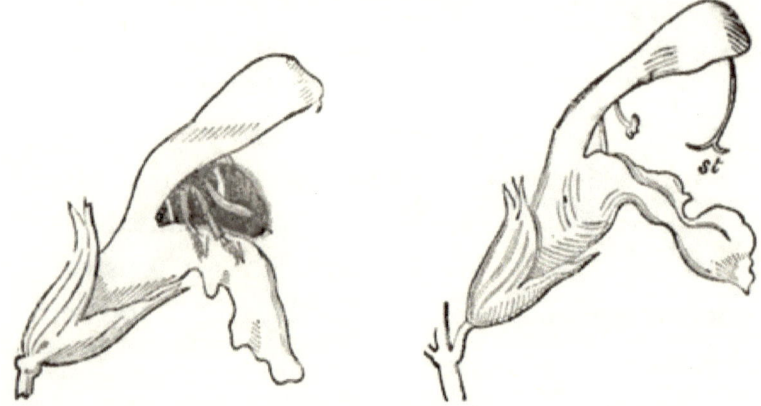

FIG. 33.—Fertilisation of Salvia. Showing bee sucking nectar, and how the connective would be disturbed so as to force down stamens on its back, just where the pistil (*st*) would be touched by the pollen.

these overgrown parts, the most important are those forming the arched hood, which protects both stamens and pistils from moisture, and the *labium* or lower enlarged petal, thrust forth for insects to light upon, as a kind of floral door-step. The latter is also frequently brightly ornamented to attract them as well. The colours of most of these Labiate flowers are blue or purplish—claimed by Dr. H. Müller as the latest evolved. But the manner in which the stamens

Fig. 34.—White Dead Nettle (*Lamium album*).

are mounted, with hinged appendages to fit them

FLORAL DIPLOMACY.

for coming down just on that part of the insect's back which would be touched by a ripe pistil, must be examined in the *Salvias*, etc., to be admired as it deserves. And the discovery that the pistil is cleft into two stigmatic surfaces, which are carefully kept together until the pollen in the same flower is all discharged, before they can open and expand without fear of self-fertilisation, will not detract from our admiration of the device. The plan of keeping the cleft stigmatic surfaces of the pistil together is one common to many other orders, although perhaps best seen in the Saxifrages and many of the Compositæ.

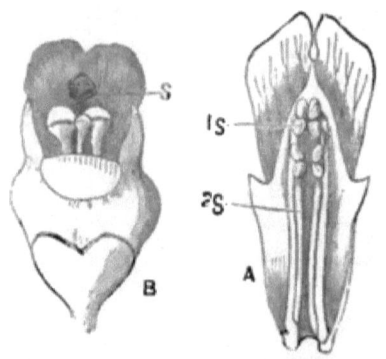

FIG. 35.—(*a*) Flower of Frog's-mouth; (*b*) Flower of Figwort or *Scrophularia*, fertilised chiefly by wasps.

The floral machinery of the Orchids is well known since the publication of Darwin's celebrated book on those plants. I refer my readers to that wonderful work for proofs of the high floral sagacity these flowers have developed. The devices of

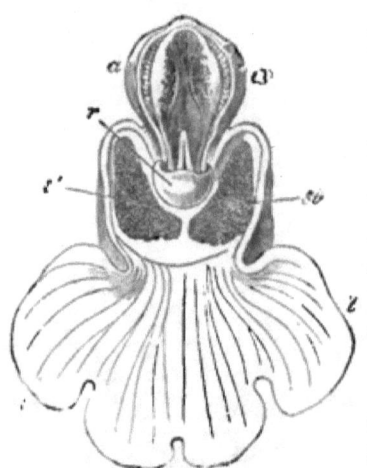

FIG. 36.—Orchid flower.

Orchids all over the world are of the most ingenious character; insomuch that Darwin's matter-of-fact investigations, as recorded in his *Various Contrivances by which Orchids are fertilised by Insects*, read like a romance. The manner in which the pollen-grains of Orchids are collected into two masses

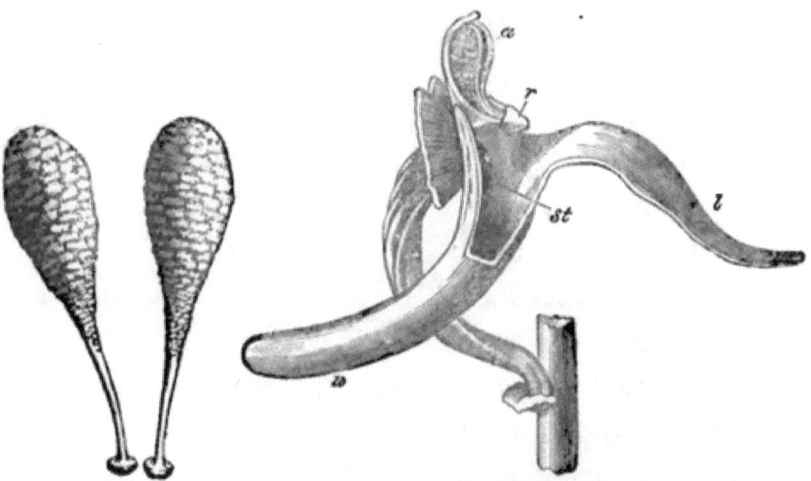

FIG. 37.—Pollen masses of Orchid.

FIG. 38.—Section of Orchid flower; (*st* stigmatic surface; (*a*) pollinia; (*r*) base of ditto.

(*pollinia*), in every species, indicates how successful this method must have proved. There is a common contrivance, adopted by species of Orchids far removed from each other by geographical distance, whereby these pollen masses are quickly detached from their little chambers when ripe, and *attached*, as by some kind of "diamond cement," to the heads of the insects visiting the flowers (a device the reader can experimentally imitate by carefully thrusting the conical point of a black-lead pencil down the throat of an Orchid flower), so that such insects carry them to other Orchids where the sticky stigmatic surfaces tear off the pollen-grains for themselves. A full knowledge of the practice increases our admiration for these aristocrats of the floral world!

The "Milkweeds" (*Asclepiadaceæ*) of America bear flowers which in some respects approach those of Orchids, both for their remarkably high organisation and structure. They proceed a stage farther than any others, except perhaps those of *Apocynum*, for they possess an apparatus which catches flies *not* adapted to fertilise their flowers, and holds them until they are dead. Two-winged flies are the greatest sufferers by this mechanism; and any one keeping Asclepiads in greenhouses will find such strangled flies in the grasp of the flowers at almost any time during the summer. We shall presently see how other plants keep away unwelcome guests by sagacious devices— the Asclepiads keep them off by real "spring-traps," ruthlessly and fatally employed, and not kept for mere threatening purposes.

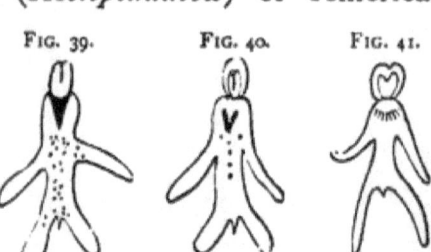

FIG. 39.—Flower of *Orchis purpurea*.
FIG. 40.—Flower of *Orchis militaris*.
FIG. 41.—Flower of *Orchis simia*.

The contrivances set up by flowers to ensure crossing are being found out almost every week, as investigation proceeds. In the works of Lubbock

FIG. 42.—Ladies' Tresses Orchid (*Spiranthes autumnalis*).

Darwin, Asa Gray, Müller, Grant Allen, and others, we have such contrivances and dodges arrayed in an almost encyclopædic form.

Tropical countries everywhere afford illustrations of flowers which seem to have of set purpose laid themselves out to attract, not insects, but *birds!* Hence the large sizes to which such flowers attain, and the extraordinary mechanism they display. The humming-birds perform their function for flowers in many of the tropical parts of America, the sun-birds, perhaps, in Southern Africa, and the brush-tongued parrakeets in Australia. If we study such structures as that of *Marcgravia* (given by Bell in his *Naturalist in Nicaragua*), we cannot fail to see how successfully its flowers have become adapted to the visits of humming-birds. Indeed, the new philosophy of botany has elevated the study to a higher platform, by showing the numberless specialities and contrivances assumed by flowers in every part of the world to secure the all-important end of cross-fertilisation. The student will find most of them in Müller's *Fertilisation of Flowers*, and will there also see the devices in vogue among a great many flowers for ensuring self-fertilisation, in case they have not succeeded in getting properly crossed. Some kinds of plants actually possess both conspicuous and inconspicuous flowers. Failures undoubtedly have been recorded, and still are experienced by these floral adventurers; but the world hears little of them, any more than it does of failure in other directions. What it does concern itself with is *success*, which rises up from the ashes of failure to a higher level than would have been attained had no such thing as failure ever occurred in the world.

CHAPTER VII.

HIDE AND SEEK.

EVEN when the numerous devices in vogue among flowers have gained the end for which all floral structures are striving—the production of seeds—there still remains the necessity for the seeds to be disseminated. The plant and its seeds are in the position of such cities as ancient Carthage, whose limited area of occupation rendered it imperative the increase of population should periodically emigrate to other quarters. Otherwise the increasing numbers would have killed each other off, by famine or pestilence, in the mutual endeavour to subsist. Nearly all plants adopt a similar policy; but, as seeds can rarely travel themselves, they are obliged to seek external aid in migrating and emigrating, and this necessity has developed devices for dispersion of the most varied kinds. When fruits and seeds are studied in this light, we are better able to understand the reason why there should be such an enormous variety, in shape, size, and colour.

Among many kinds of fungi, water-weeds, sea-weeds, mosses, and even ferns, the spores and male organs actually possess locomotive power, and by means of cilia and flagella are able to move from the parent plant, and distribute themselves to some distance. The names of *zoospores, antherozoa*, etc., have been given to these organs of reproduction. The spores of other kinds of fungi, as well as those of mosses, ferns, etc., are of exceeding small size, even in the largest species; and they are produced in amazing numbers. Their smallness and lightness are in their favour, for the wind readily disperses them, and carries them to great distances, perhaps even across seas and oceans.

There is a difference between the botanical and the popular idea of the fruit. The latter regards it as "something to eat!" The former considers it as the ripened pistil, or seed-case. Ruskin's description aptly places the relation of seed and fruit before us—the fruit is in reality the "husk." This husk may be eatable—it often is, as in the plum, apricot, peach, etc., but it is often, perhaps oftener, not edible. Nay, the edible part may not be a husk at all, although it exists for the sake of the husk, as in the strawberry.

One has not to consider very long before arriving at the conclusion that true fruits, like true flowers, may be divided into two classes—conspicuous and inconspicuous. There is this difference in the parallel, however—fruits have undergone perhaps more modifications (seeing how important they are to the perpetuation of the species) than even flowers. But just as conspicuous and attractive flowers owe their qualities to the necessity for *insect* visits, and the crossing consequent upon them, so have conspicuous and edible fruits been evolved chiefly through the agency of *birds*, to whose appetites they appealed in the first instance!

It is both curious and interesting to find that the parts which protect, conceal, and defend the seeds of some plants, are those which have been specially modified and adapted to attract attention in others; and that both sets of devices have been evolved for the sake of benefiting the seeds. Thus the external parts of the plum, cherry, peach, and other fruits, are botanically the same covering as the hard woody "shells" of the hazel and other nuts. In the former instance the pericarp has grown sweet and succulent, and has become brightly coloured and attractive as well, so that it might catch the eyes of birds. Fruit-eating birds have gradually become specialised to this habit, as fruits have been developed. That the latter have slowly caused their pericarps to grow sweet and pulpy is proved by the ease with which cultivation has, within the historic period, taken up the matter, and rapidly converted our small wild fruits (big enough, however, even in their natural state, to suit the mouths and stomachs of birds) into the large and succulent varieties which now take the prizes at horticultural shows. *Natural* selection commenced the task, and *artificial* selection has completed it.

In the summer time the popping of guns in the orchards tells us the birds are now being kept away. Thrushes, blackbirds, and other fruit-loving kinds are falling victims, or being driven away from the cherries, currants, and strawberries. Few people are aware that if it had not been for them and their ancestors, we should never have had the very fruits from which we now frighten them away. In this way man has "annexed" the long-laboured results of frugivorous birds. Originally the fruits were just large enough for such birds to swallow them separately—now they have been developed by artificial selection to such a size that our fruit-eating birds cannot bolt them whole, but eat away the thick layer of succulent pericarp, leaving the stones attached to the stalks!

Had plums, cherries, peaches, and apricots, not been favoured by man, increase of bulk and size would have been a serious matter for such fruits. For if the stones of these fruits had been left hanging to the stalks the seeds could not have been dispersed; the entire purpose for which the juicy pericarp was originally developed would have been counteracted, and such species of plants would have slowly died out. Man has taken the matter of their care and dispersion under his own management. Fruits have learned, so to speak, that men are more profitable to them than birds, and that they have now a much better chance of survival and dispersal by human than by ornithological agency; and so they have eagerly responded to his horticultural labours, and grown as large and sweet and palatable to his taste as possible. Still, one cannot help thinking how badly their first love, the birds, fares at the hands of their second love, man!

But enormous numbers of fruits still retain the average size which originally plums, cherries, etc., possessed. This is indicated by the size of the wild bird cherry, from which at least one group of the cultivated kind has been horticulturally developed. The common sloe of our hedgerows is of about the same bulk—that of a pea; although some of our largest plums have been cultivated from a southern natural variety of the sloe. This average-size of fruit is still maintained by numerous wild species, as those of the Haw, Mountain Ash, Service-tree, Yew, Honeysuckle, Bryony, Mistletoe, Bitter-sweet, etc. The fruits of the Cranberry, Crowberry, Bearberry, etc., are rather smaller; whilst those of the Raspberry, Blackberry, Dewberry, and Cloudberry are in reality *clusters* of smaller fruits whose individual structure is identical with that of a plum or cherry. So that the birds can pick off any one of these compound fruits at will, and swallow as many of the little drupels as they like.

All the above-mentioned fruits, whose attractive pulpy exteriors induce birds greedily to swallow them, possess hard stones, which resist the action of digestion in the bird's stomach, and effectually protect the kernel or seed inside from any chemical

Fig. 43.—Scarlet Bearberry (*Arctostaphylos uva-ursii*).

Fig. 44.—Crowberry (*Empetrum nigrum*).

decomposition. Thus, after birds have made a hearty meal off the particular kinds of berries they are fond of, they fly away perhaps to great distances, and scatter the undigested stones and seeds in their droppings. Such seeds cannot fly and disperse themselves, like those of the thistle, but they cunningly induce the birds to be their carriers, and they honestly pay them for their trouble!

When these seeds eventually fall to the ground they are at such a distance from the parent-plant that there is no danger of overcrowding. Moreover, they find themselves surrounded by a sufficiency of fertile manure supplied by their carriers, which cannot but prove favourable to their germination.

In such of the smaller fruits as have no hard stones there is usually an inner tough and horny layer which serves quite as effectually to protect the minute seed within from harm. This may be seen in the juicy little fruits of the Dewberry, Blackberry, and Raspberry. Can we wonder our hedgerows should be so abundantly upholstered with many varieties of Bramble when we remember the annual feasts of blackberries they offer to their best disseminators, the birds, and that each blackberry is, in reality, a compound fruit, containing twenty or thirty seeds?

This simple and largely-practised device of inducing birds to disseminate seeds by altering the usually hard, tough, and unattractive husk so that it becomes a delicate and choice morsel, is not confined to any particular order of plants, although that of the Roses has perhaps most largely adopted it. But this latter order, so remarkable for the members of its species being of the kind we call "fruit-bearing," includes some which have preferred other methods of attracting birds. Among such are the Strawberry, Rose-hips, Mountain Ash, Service-tree, Apples, Pears, etc., most of which have now, like Plums and Cherries, been transferred from the agency of birds to that of man, under whose care they have grown larger, more succulent, and at the same time developed new flavours and odours of which they knew nothing in their lower ornithological condition.

Of these several devices, originally put forth to attract birds, and be a favourite food with them, none are more ingenious than that adopted by the strawberry. It seems almost a libel upon this most delicious morsel to say it is not a fruit—but the truth must out! All of that part for which we value it is the *receptacle*—the same part on which the combined fruits of the raspberry fit, like a thimble on the end of the finger. Nobody thinks of eating it in the raspberry; few people would attempt to do so twice. When the florets of the Thistle have all been pulled off, a similar but flattish green receptacle is seen, which country boys will sometimes eat—but what will not boys attempt in this direction! To compare even the wild strawberry with such a tasteless representative, however, indicates what a wonderful change must have taken place at some period or another in the history of the Strawberry. Of course the cultivated kinds are here left out of the comparison.

The fact is, the receptacle of the Strawberry has

become juicy to induce birds to swallow it. In its natural state its size is about the same as that of the real wild fruits above mentioned. The so-called

FIG. 45.—Strawberry. *f*, shows positions of true fruits on the globular succulent receptacle.

"seeds" scattered all over the surface of the strawberry are in reality *genuine fruits*, consisting of husk and seed. Each is a true nut, although of almost microscopic size, with the kernel or real seed within. The husks are horny, and admirably adapted to protect their charge; and both pass out of the bodies of birds in practically an unaltered condition. Even their smallness is of advantage to them, in enabling large numbers to be swallowed at a single mouthful, just as children are induced to swallow nauseous pills in the spoonful of jam which masks them!

The sweet part in a dried fig is similar to the pulpy part of the strawberry. It is a receptacle turned outside in, and all the so-called "seeds" of the fig, minute though they are, are true botanical fruits, comprising husk and seed. As in the strawberry, these numerous fruits of the fig gain by their smallness.

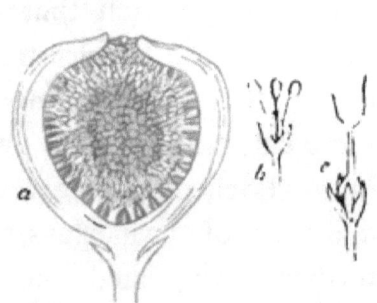

FIG. 46.—Section of fig; *a*, receptacle; *b*, male flower; *c*, female flower.

The hips of our hedgerows are a more elaborate

botanical contrivance; although their general mechanical structure reminds us of the fig, for the bright red, soft, and somewhat sweet objects we see so conspicuously in the winter time are the altered calyx-tubes. The hairy objects within are fruits; each contains a seed; and when birds devour the succulent hip (which does not properly ripen until wintry weather has been experienced, and the insect and other food of birds is therefore scarcer), they consume the achenes within, or otherwise carry them off adhering by their hairy exteriors to the sides of their mouths—anyhow, to be dropped at a distance from the parent shrub.

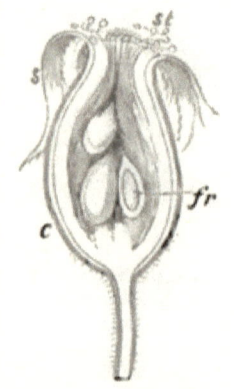

FIG. 47.—Rose fruit or hip; *c*, scarlet calyx-tube; *fr*, true fruits inside.

Apples, Pears, and Medlars had best be studied in the wild state, or in those species which are believed to approach that condition, as in the Crab-Apple and the Service-tree. Then we shall see that our modern fruits are merely a fleshy growth of the same calyx-tubes, until eventually they have become quite attached to the ovary. Of course, in the apples and pears of our gardens, their fleshy growth has been further exaggerated and developed, and its quality and flavour improved. But when one of these fruits is cut across, we see the horny carpels of the ovary containing the seeds or "pippins." What a contrast in the behaviour of the calyx in these fruits and in those of the thistle! In apples and pears it grows fleshy and bulky—in the latter it gets drier and lighter. But in both instances the same end is subserved for the seeds; the former enables them to be disseminated by birds and mammals, and the feathery parachute or "pappus" of the latter is ingeniously contrived so that the seeds can be dispersed by the wind!

I have drawn attention to the vast number of plants which have selected bird-agency for scattering their seeds. The fact that members of various orders have laid themselves out for this purpose, and have hit upon similar or analogous contrivances, in so many places far removed from each other, is a pregnant fact to the philosophical botanist. It cannot be understood without calling in the aid which the modern doctrine of evolution affords. Black and red appear to be the chief colours adopted by fruits, although some have an attractive bluish "bloom" about them, as in the sloe, grapes, etc. The berries of the Mistletoe are of a whitish hue. But black (a very rare colour in the petals of flowers) is common among bird-distributed fruits. In the same order, and not unfrequently even in the same family, as among the *Ericaceæ*, for instance, we find the most opposite results—some producing succulent fruits, like the cranberry and whinberry; and others dry ones, as the Common Heath, etc. Next to the Roses, this order has been most successful among our British plants in evolving pleasant and edible fruits.

Externally, nothing can be more unlike than the dry cones of a Fir or Pine-tree, and the scarlet, succulent fruits of the Yew, or the bloom-covered blackish fruits of the Juniper. And yet, if we carefully observe the early stages of their development, we shall find that all commence alike as true cones. In the Yew and Juniper some bracts become aborted, and others unusually developed, until their well-known fruits are the result.

In the Mulberry, again, we have a bird-dispersed fruit produced by the bracts of the hanging cluster of flowers becoming fleshy and sweet, and growing so large that eventually they get fused together into one mass, when they look like unusually big blackberries. It is to this series of modifications that the fruit we call the "mulberry" owes its origin.

In our British flora, the following are among the commonest of bird-disseminated fruits:—Strawberry, Blackberry, Dewberry, Cloudberry, Barberry, Arbutus, Privet, Spindle-tree, Guelder-Rose, Buckthorn, Holly, Ivy, Honeysuckle, Bryony, Yew, Mistletoe, Sloe, Arum, Bird-Cherry, Haw, Wild Rose, Mountain Ash, White Beam-tree, Wild-Service tree, Crab-tree, Bearberry, Baneberry, Blaeberry, Juniper, Crowberry, Cowberry, Cranberry, Asparagus, Butcher's Broom, etc.,—a long, varied, and interesting list, speaking volumes for the changes which have collected so various an assemblage into one common group, as regards the means adopted for the dispersal of their seeds.

Man has eagerly seized upon those species of succulent fruits which happened to suit his taste and appetite, and has practically robbed the birds of them. But there still remain many kinds he is in no hurry to "annex." Some are regarded by him as *poisonous*—many of them are *nauseous*, as the attractive berries of the Honeysuckle, Bitter-sweet, Bryony, and Yew, for instance. Those of the Juniper are employed in the manufacture of gin, but they have their revenge on man in another way.

The fact is, many of these fruits have discovered they were liable to be devoured by animals unserviceable to them for the purposes of dissemination, and so they have gradually secreted in their pulpy flesh substances objectionable to mammals, but to which birds have grown accustomed to consume with the utmost safety. Such are the fruits of the Bryony, Bitter-sweet, Black Nightshade, Yew, and perhaps those of the Arum and Deadly Nightshade (*Atropa belladonna*) as well. In South America, we read that one of the most fatal of all known poisonous fruits, the Manchineel (*Hippomane mancinella*), is eaten with impunity by native birds. The fruits which have thus learned to confine themselves to birds are evidently gainers by the artifice, and we can therefore understand why many of them, in this respect, advertise that "no mammals need apply." Usually, plants have coloured pericarps only, but a few have this part uncoloured, and their seeds of the most vivid hues instead. Among the latter are those which the country people called the "Roast-beef" plant (*Iris fœtidissima*). In its case the hard seeds, when swallowed by birds, are not digested, and perhaps the agency only of very young and inexperienced birds is employed in this operation, useful to the plant, but hardly paying the birds for their trouble. Indeed, compared with the honest methods adopted by the Plum, Cherry, etc., of rewarding the bird-carriers of their seeds with a good meal, the *Irises* get theirs dispersed "under false pretences," for the young birds pick up their scarlet seeds, misled by the instinct of their race, which teaches them that attractive fruits are good to eat, and then find out they have been duped into partaking of what to them are equivalent to the "apples of Sodom!"

The Spindle-tree (*Euonymus europæus*), abundant in the hedgerows of eastern and southern England, has advanced so far along the path of fruit-differentiation that both its arillus and enclosed seeds are attractively coloured. The former is of a bright crimson, and the latter of quite as vivid a scarlet. Moreover, the carpels open in the autumn time, when the leaves have nearly all fallen, and the fruit is therefore better exposed. The four scarlet seeds remain attached to the four expanded crimson coverings, so as to render the Spindle-tree fruit the most attractive perhaps of our native species. These fruits are so beautiful that the branches bearing them are frequently plucked and placed in vases for the sake of their attractive and flower-like appearance. The birds, however, do not allow them to hang long —plain evidence of the success of their chromatic device.

The elaborate and successful method adopted by the fruits of our Spindle-tree, of displaying *two* colours instead of only one, as is usually the case, is practised elsewhere. In the West Indian Islands there is a species of *Pithecolobium* whose curious pods curl two or three times round, like those of the Medic, and when they split, they are bright red inside, which acts as a foil to the bright-blue seeds.

In the South Pacific Islands the natives utilise some of the bright-coloured seeds, and convert them into neck-beads, with excellent effect. One point is worth noting in all these brightly-coloured seeds— they are usually *scarlet*, and always remarkably hard. Consequently they take no harm from being soaked a few hours in an animal's stomach; for their superior hardness protects them from digestion, and afterwards they are disseminated successfully.

The larger fruits of tropical and equatorial regions have been evolved side by side, and contemporaneously with the larger and more brilliant kinds of fruit-eating birds peculiar to those climates. Mr. Grant Allen and others have pointed out the association of bright colours in butterflies and birds generally, with the attractive flowers and fruits they feed upon. These surroundings may have led to sexual selection operating in certain directions; on the principle of the old proverb, that we can tell a man's character by the company he keeps.

There are not many species of fruits in Great Britain which lay themselves out for dissemination by the aid of mammals; but representatives are not wanting. In some countries the number of such fruits is oftimes great, and the mechanism developed for the purpose of conveyance very ingenious. The Burdock (*Arctium lappa*), found in most waste places —a plant which has run into several well-marked varieties—is perhaps the most striking. It owes its popular name to the "burs" or flower-heads, set all round with numerous hooks which take hold of the clothing of a man, or the hairy hide of a passing animal, with the greatest facility. The seed-vessels or fruits of the Goose-grass or "Cleavers" (*Galium aparine*) are also crowded over with a similar minute fish-hook-like mechanism with which both its stem and fruits are covered.

In countries where wild animals are more numerous, plants bearing mammal-dispersed seeds are more plentiful, and their mechanical adaptation to this kind of dissemination is shown by the number of fruits and seeds, which are brought over to this country and France, in the foreign fleeces imported by wool merchants. In some of the Gloucestershire valleys, where foreign wool is washed and worked, we find plants from the Cape or South America germinating and sometimes flowering, the seeds having been brought over in the wool. Most of them are hooked, and some very formidably, such as *Xanthium strumarium* and *X. spinosum*, *Martynia*, etc. At Montpellier, in France, numerous seeds brought over in fleeces from Buenos Ayres and Mexico have sprung up, insomuch that botanists have been enabled so far to study the flora of those countries, thus imported in an unexpected but effective way into Europe across the Atlantic. In Africa and Madagascar we have plants whose fruits are so well fitted for this kind of conveyance that they go by the name of "grapnel plants" (*Harpagophytum*); and in South America the *Proboscidea jussieui* affords perhaps the best example of such a kind of fruit-mechanism in the New World.

A few species of plants, not confined to any particular order, have invented another scheme for dissemination, which may be called "mechanical." One of the most remarkable of these is the well-known "Squirting Cucumber" (*Momordica elaterium*), whose ripened carpels have to be held together by copper wire if it is desired to keep them together. Dr. M. C. Cooke mentions the North American Witch Hazel (*Hammelis virginica*) as being elastic enough to strike passers-by violently with its expelled seeds. Our Common Balsam (*Impatiens-noli-me-tangere*) scatters its seeds in a similar manner; and those grown so commonly in cottage gardens perhaps owe their cultivation to the love of their owners for a bit of harmless practical joking—which consists in getting an uninitiated person to just touch the ripe fruits, and his sudden startle upon the thing's "going off" is a sufficient reward! On a very hot day in summer, where the Gorse-bushes abound, we hear quite a fusillade, caused by the ripe pods suddenly opening and expelling the seeds. Dr. Cooke mentions certain kinds of fungi which eject their spores by similar methods.

But the wind, all the world over, in spite of its proverbial fickleness, is largely depended upon for the dissemination of many kinds of seeds, just as we have seen it is utilised for cross-fertilising many kinds of flowers adapted for that purpose. The specialisation of seeds for wind-dissemination is often a very elaborate matter. One particular order of plants seems to have laid itself out for wind agency to distribute its seeds—the *Compositæ*. How successfully the trick has answered is seen in the cosmopolitan distribution of the members of this order, the fact that it stands at the head of all others in numbers both of species and individuals, and the marvellous manner with which its members have adapted themselves to almost any physical condition on the earth's surface. Arctic, antarctic, tropical, and equatorial regions alike include them among their chief plants; the *Compositæ* grow in arid regions like the Cape Bush (where they vindicate their vitality by blossoming perhaps in their very brightest and loveliest tints as " Everlastings "); on the dank tropical hillsides of Brazil; in salt marshes, bogs, rich alluvial soils, barren rocks, desert, and plain. They assume both herbaceous and woody or arborescent forms as occasion requires. No place is too hot or cold, high or low, wet or dry, for the species of this elastic order. They are the " Scotchmen " among the vegetable world—found everywhere, and making a good honest living wherever found!

Of course, all the members of the *Compositæ* do not resort to wind-dispersion; but it is remarkable that it should be so largely employed by them. As has already been stated, the mechanism used for the purpose is the *pappus*, or " clock," best known in the Dandelion. This structure is well illustrated by a child's shuttlecock. In the Goat's-beard (*Tragopogon pratensis*, etc.) the rounded and undisturbed " clock " is a most beautiful object. Everybody is acquainted with it in the Dandelion and Thistles. The slightest breeze lifts one of the feathery pappi and carries it away, with its fruit dry and light for the purpose hanging beneath, and balancing it like the car of a balloon. This structure has been evidently evolved from several standpoints, as the different finished examples of it show. Sometimes the "clock" or pappus has a long stalk, as in the Dandelion; and at others it is quite without, as in the Ragwort (*Senecio jacobæa*). Even the cottony hairs of the "clock" have been improved upon by the Goat's-beard (*Tragopogon pratensis*), the Artichoke, and Carline Thistle (*Carlina vulgaris*), etc., in which they have become *plumose* or feathery—a complexer structure, and indubitably one better fitted to be carried by the wind, for the feathery hairs interlock into a hollow cone like the selected feathers in the more expensive kinds of children's shuttlecocks.

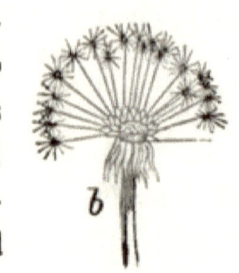

FIG. 48. — Dandelion — with feathery pappi attached to fruits; *b*, Ripe flower-head.

Singularly enough, in some mammal-dispersed fruits, the hooks they bear are formed by parts of the calyx being persistent, and becoming hard and woody. Perhaps no other auxiliary floral organ is so ready to turn itself to advantage, and be a "Jack-of-all-trades" to the valuable seeds it subserves, as the calyx. The better we are acquainted with structural botany, the more are we surprised at the numerous characters assumed by this organ. They are always developed in view of some advantage to the seeds and their dissemination.

Those near relations to the Thistles, and therefore to the *Compositæ*—the Teasels (*Dipsaceæ*) have managed to take out the same patent as their first cousins, and to modify the free portion of their calyx into a "clock" that the wind may waft their fruits away. Singularly enough, another order of plants far removed from both the above, the *Valerianaceæ* have adopted a very similar contrivance, for the same purpose. And one species, the Red Valerian (*Centranthus ruber*), abundant in most cottage gardens, has even succeeded in developing a pretty and highly-elaborate feathered *pappus*.

The Cotton-plant (*Gossypium*) is now clothing more than half the human race with the long cellulose hairs with which the surfaces of the seeds were originally covered for the purpose of enabling the wind to scatter them, just as we see it dispersing the cotton-covered seeds of the Poplar, Cotton-grass (*Eriophorum*), and Sedges in our own country. Here the end is gained by quite a different contrivance to that adopted by the Dandelion and Thistle.

Sometimes the services of the wind are enlisted to convey seeds to a distance from the parent-plant, not by ballooning, but on the screw-propeller principle. This is illustrated by those peculiar kinds of fruits called *Samaras*, well known in the Maple, Ash, Elm, and Birch. This type of mechanism is generally borne by trees, and seldom by shrubs, and perhaps never by herbaceous plants. For the greater height of trees assists in the work, and when a *samara* is detached the winged expansions are caught by the wind, and revolved like the blades of a screw-propeller, so that the seeds are thus carried to some distance. In the Lime (*Tilia Europæa*) the fruit is borne on a long stalk, to which a papery bract grows for about half way. The latter gets drier as the fruit ripens, and then bends back in such a way that it is whirled to a distance, just as if the fruit of the Lime had winged expansions or special *samaras* for the purpose; and when the time for their dispersion comes we see them flying about in the air like so many butterflies.

Everywhere, wind-agency has been evoked by plants for the dispersion of their seeds, and in some way or another has been enlisted in their service.

Not a few plants trust to even more accidental means of transport—to the currents of the sea (as the Cocoa-nuts, etc.), to streams and rivers; and these fruits can remain in salt or fresh water, as the case may be, for an unusually long time without suffering harm; whereas fruits and seeds not adapted to dispersion by such means soon die and rot in the water. Even the varying degrees of vitality possessed by seeds are related to their competition for life. Some have been found in Celtic tumuli, like Dr. Lindley's Raspberry seeds, sown in Chiswick Gardens after an interval of perhaps two thousand years, and then germinating into the plants whose descendants, I believe, are still to be seen there! Similar successful experiments have been made with seeds found in ancient Roman tombs. And everybody knows, how, when a wood or a heath has been burned down, and all the recent plants destroyed, an entirely new set of plants immediately shoots up—their seeds having been quietly waiting in the earth for some such chance as this—a chance which was materially affected by their power to retain vitality for an indefinitely long period of time.

CHAPTER VIII.

"DEFENCE, NOT DEFIANCE."

This well-known motto of our volunteers has long been adopted as a vital principle in the vegetable kingdom, and numerous plant peculiarities can only be properly understood when we bear this in mind. True, some of the lower forms of plant-life have assumed a parasitic life even upon animals, such as the Bacteria, etc., and may consequently be regarded as predaceous. The so-called "carnivorous plants," also, have turned the tables upon the ancient enemies of their kind; but they are quite exceptional in their structures and habits.

All animals, except the purely carnivorous kinds, live upon plants. Even carnivorous and insectivorous animals and birds are dependent on their prey feeding upon vegetation. It is a marvel how plants have been able to withstand this universal and continued assault—root, stem, leaves, and fruit, all are partaken of. Individual growth and specific propagation are interfered with at every stage of their development. To attack them insect forms have been specialised in a thousand ways, and they carry on their assaults in the most ingenious manner and with the greatest hardihood. Birds and mammals in countless myriads depend upon the growth of the vegetable kingdom for their daily bread.

When we remember the probability that vegetable life preceded animal life in its appearance upon the globe—that from lowly organised forms plants have gradually been evolved to more complex and highly organised types; that in genera, and species, and number of individuals, they have increased and multiplied, so that when man appeared they had attained their maximum development—our cause for wonder at the success which has outlived the attacks of the animal kingdom, nay, which has even turned its members to account, and pressed them into its own service for the fertilisation of its flowers and the dissemination of its seeds, is not decreased!

A good many people who have not studied plants, and who still hold the comfortable and old-fashioned doctrine—unquestioned until a few years ago—that all things were created for the use and service of man, cannot understand why all plants should not be equally useful to him. They have taken refuge either in the idea that all plants do actually possess some good qualities, if we only knew them; or else that the presence of poisonous and baneful plants is attributable to such a catastrophe as the " Fall of Man ! "

Such notions cannot be compared with the larger views of the Creator's power and wisdom with which modern science is replacing the venerated, but in reality crude and even *materialistic*, opinions of the last century.

Having seen that plants, as a rule, are defensive and seldom predatory objects, we can better understand that their continued existence and vegetable triumph depends upon the strength of their defences. These are of innumerable kinds ; but the easiest and most effective method is the secretion of some *poisonous* or deterrent principle by the organs most liable to attack, or which are of the greatest importance in the economy of the plant.

Take the secretion of *poisons*, for instance,—the faculty for accumulating them has brought many plants within the scope of superstitious speculation. Vegetable poisons are neither peculiarly the property of any one order, nor even of a species. Some plants have succeeded in developing them in a much higher degree than others, as the Manihot, Manchineel, the Sumachs, etc. Certain orders of plants have even attained a high degree of security and immunity from attack by the secretion of strong poison, as the *Euphorbiaceæ*, represented by such intensely poisonous species as our Common Spurges, the *Euphorbia virosa*, etc., of Africa, and *E. cotinifolia* in Brazil.

The Solanaceæ is also a highly defended order of plants in this respect—few species being free from poisonous or other secretion deleterious to animals; and one of them, the Belladonna (*Atropa belladonna*, generically well named after that individual of the "Fates" who cut the thread of human life), has attained a high notoriety for its poisonous character. The Bitter-sweet (*Solanum dulcamara*) and the Black Nightshade (*Solanum nigrum*) are doubtfully poisonous, but certainly to other animals than birds their berries are not wholesome; and the fruits of the Potato (*Solanum tuberosum*) are so objectionable that few kinds of animals partake of them—thus furnishing a marvellous contrast to another species of this order, the Tomato (*Lycopersicum esculentum*), whose beautiful "love-apples" bid for the services of animals to carry away the hard seeds!

Another genus of this order, known in the West Indies as the Poison-berry (*Cestrum*) sufficiently denotes by its popular name its deterrent qualities. It might be objected that if animals partook of such vegetable poisons there would be an end to the matter, for both attacked and attackers would become extinct together. But animals which have eaten of poisonous seeds frequently recover, and the lesson thus learned is not only never forgotten, but the experience becomes a racial tradition or instinct. Moreover, it is the young rather than the old birds and mammals which make such a mistake. When cattle were first conveyed to South America, the Cape, and Australia, many died through eating poisonous herbs. Their descendants, however, have by this time learned which are safe and good to eat and which are not, and those which fall victims are usually young animals. Thus natural selection weeds out the unwary.

Perhaps it would be found that, in highly-organised plants, those parts which are usually most poisonous are the fruits and seeds. The former is really the "seed-case," and it is, therefore, a higher and completer protection for it to be *bitter* to the taste or otherwise deterrent, for then animals might proceed no further, and the seeds would be effectually protected. Some seeds, like the Nux-vomica (*Strychnos nux-vomica*), for instance, have acquired a most virulent poisoning power. The *Strychnos* is a genus remarkable for poisonous secretions. Sometimes, as in *Strychnos ticuté* (a Javanese climbing species), juices are produced poisonous enough for the natives to dip their arrows in. Another species, *Strychnos toxifera*, (rightly so called), well known as "Wourah," is one of the most frightful poisons in the world. *Strychnos colubrina*, is powerful enough to be an antidote against snake-bites.

Poisonous fruits and seeds are generally produced at some distance from the earth, and therefore birds are the chief animals which partake of them, unless they fall to the ground. We have seen how birds have been converted into useful friends, to disperse the seeds, instead of being constantly contended with as enemies. Moreover, birds are so constituted that fruits poisonous to other animals are probably not so to them. They are, as a race, in the condition of an arsenic or opium-eater, who, by constantly partaking of such drugs, is by and by enabled to safely consume what would be a fatal dose to other persons. In this way such poisonous fruits as yew berries are protected from mammals (which would be unserviceable and even injurious to the seeds), and are confined to birds which have become constituted to partake of them without injury. I believe that the bright red fruits of the Lords and Ladies (*Arum maculatum*), which stand forth in prominent clusters in our country lanes in the autumn time, are not guilty of the deed ascribed to them by Grant Allen, viz.—of attracting birds to eat them, and then poisoning

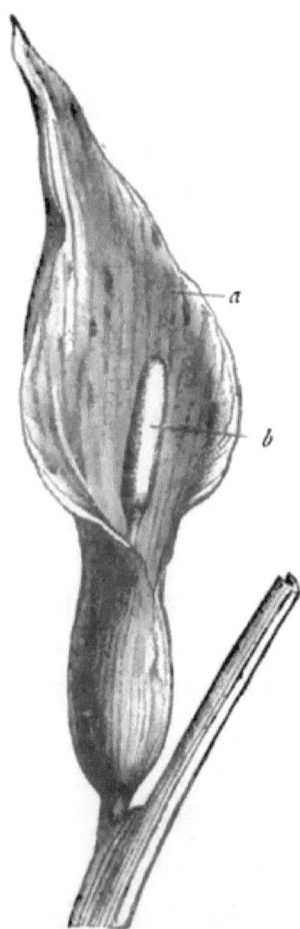

FIG. 49.—Arum, or Cuckoo Pint. (*a*) spathe; (*b*) spadix.

them, so as to convert their carcases into a rich manure heap! Had such been the case, birds would have long ago got used to the device, and have subvented it, either by being inured to the poison or by not partaking of the fruit. But the abundance of the Arum shows that its berries are not poisonous to some birds, although they may be to other animals, for thrushes partake of them with safety, although they do not resort to the berries when an abundance of more favourite food is at hand.

The efflorescence of wax known by the name of "bloom" on fruits may perhaps be regarded as a protective device, originated for the special purpose of guarding such fruits from decomposition. Rain and dew roll off the surfaces of the fruits covered by "bloom," when otherwise the moisture would not only rot the fruits, but might possibly deposit upon them the germs of microscopic fungi, etc. The general absence of minute vegetable parasites upon them, and the length of time ripe grapes, plums, etc., will hang uninjured if their "bloom" is not interfered with, shows how effective this special secretion has proved. Possibly the deposit also prevents the evaporation of the internal juices, and so keeps them looking plump and attractive for a longer period of time than otherwise would be the case.

Poisons are frequently secreted in the leaves, bark, and other tissues of plants; but some other chemical deterrent principle is generally employed, such as oxalic acid, tannin, etc. Singularly enough these objectionable substances are found most abundant in adult plants, so that they must have reference to flowering and seeding. Moreover, they also indicate that such poisonous defences have been acquired. Cattle often partake of objectionable *young* plants when they will not eat them, even if pressed by hunger, in the adult or matured state.

Plants growing on or near the ground frequently employ some kind of acrid poison by way of protection, like that secreted in the leaves of the "Lords and Ladies," for instance. The *acridity* frequently found in the leaves and stems of soft herbaceous plants is perhaps the commonest form of their defences. It is present in varying degrees up to a virulent poison in the *Ranunculaceæ*. One British species, the Celery-leaved Buttercup (*Ranunculus sceleratus*), is so intensely acrid that half an ounce of its juice will kill a dog; and Professor Oliver mentions that the root of an Indian species of this order, *Anemone ferox*, affords one of the Bikh poisons used in the Himalaya to poison arrows for tiger-shooting. Some of the most virulent poisons employed in medicine are extracted from this order, which, all over the world, has somehow managed to take the lead in this respect among herbaceous plants. Not a few species show by their common names the uses to which they have been put, as the Wolf's-bane (*Aconitum*). How effectual a protection to the leaves this acrid secretion is may be seen in any well-cropped field in the summer time. It is yellow with abundance of Buttercup flowers, and thousands of plants are distributed, growing rank and healthy amid the well-grazed grass, but all untouched by cattle. Sheep will feed upon them when young; and the tougher and less dainty goat manages to make a meal of the common Buttercup (*Ranunculus acris*); but neither horses nor cows will have anything to do with it. There is a semi-humorous aspect in the teleology of this poison secretion, for tramps are in the habit of using the leaves of the last-mentioned species of Buttercup, as well as those of the Celery-leaved kind (*R. sceleratus*), to produce blisters on their limbs, in order to excite compassion in the hearts of the unitiated, and extract the coin with which a well-to-do Briton likes to relieve himself from the distress of a pitiful sight!

Mr. Gosse describes the Dumb-cane (*Caladium seguinum*) as growing in rank abundance in certain damp dells in Jamaica. It is a tall kind of Arum, "so virulently acrid that the juice of any part incautiously applied to the mouth causes the tongue to swell so as to take away the power of speech, and produces burning torments of long duration. It is said to have been one of the modes of torture employed by cruel masters in the dark days of slavery." Our familiar and abundant English species, *Arum maculatum*, possesses nearly as intense an acridity ; and one of the brutal bits of practical joking formerly in vogue consisted in inducing people to chew an Arum leaf under the idea that it was Sorrel. The pain resulting was both intense and long-lived. But in consequence of this acridity the *Arum* enjoys immunity both from mammals and caterpillars, and flaunts its glossy green leaves boldly forth in defiance of both these kinds of hereditary enemies.

A reference to any book on practical botany will show the reader that the leaves of many other kinds of common plants are not eaten because they are disagreeable or nauseous, rather than actually poisonous. This is the case with many of the *Cruciferæ*, the most harmless order of plants, and therefore in striking contrast with the Buttercup family (*Ranunculaceæ*). Among orders whose members have struck out in search of different protective substances, the most remarkable, perhaps, is the *Scrophulariaceæ*, which includes such actually poisonous members as the Foxglove (*Digitalis purpurea*); others, like the Figwort (*Scrophularia nodosa*), which develop such a rank smell, and possess such a bitter taste, that only goats are known to eat it, at least among mammals; whilst the Germander Speedwell (*Veronica chamædrys*), loveliest of British wayside flowers, secrete so much astringency that their leaves are protected in a surprising manner. The Mulleins (*Verbascum*) adopt another plan, and cover their leaves with abundant woolly hairs, which must be as tempting to cattle as offering them strips of flannel!

Many kinds of the fungi popularly known as

Fig. 50.—Wood Sorrel (*Oxalis acetosella*).

"Toadstools" are poisonous, or otherwise so unpalatable that they are never eaten except by certain slugs and the larvæ of beetles, whose omnivorous appetites have enabled them to withstand what would be poisonous to other animals. In some species of fungus, like the Fly Agaric (*Amanita muscarius*), the poison is intense. In the leaves of Wood Sorrel (*Oxalis acetosella*) and the Common Sorrel (*Rumex acetosa*) a well-known poison, oxalate of potash, is present, and so we find luxuriant clusters of the green leaves of the latter in all rich pastures quite untouched; whilst those of the former plant frequently monopolise the ground in damp but not too dark woods.

Citric and malic acids are neither of them liked by mammals or birds, and they are usually present in great abundance when such fruits as plums, apricots, cherries, gooseberries, apples, pears, strawberries, cranberries, raspberries, blackberries, etc., are young and unripe. It is necessary they should be protected in this immature state, or else they might be devoured before their enclosed seeds were ripe and ready for dispersion. The policy of keeping off too greedy friends by the temporary secretion of acids they don't like has therefore proved very effectual.

One order of widely-spread plants, the *Gentianaceæ*, is remarkable in that nearly all its species have an intense bitterness. How largely this protects them is shown by their being seldom eaten, even by caterpillars. This bitter principle is common to genera included in the order, of otherwise extremely varying character, such as the Centaury (*Erythrium centaureum*), the Yellow-wort (*Chlora perfoliata*), the

Bog-bean (*Menyanthes trifoliata*); as well as the whole genus of Gentians, Alpine and otherwise.

Kerner has shown that the mechanism adopted by some plants, such as *Chlora perfoliata*, the Honeysuckle, Teazel, etc., of having the bases of their opposite leaves *connate*, or growing together and embracing the stem, must prevent creeping insects from getting to the nectar of the flowers. We may frequently see numbers of small dead flies in the pools of rain and dew collected in the connate leaves of the Teazel. Those of the Butter-wort (*Pinguicula vulgaris*) have a ridge raised all round the margin which causes them to hold moisture, and the flower-stalk is therefore as much isolated as if it actually stood in water, according to the similar device adopted in some countries of placing the table-legs to prevent insects from crawling up the table.

The *Bulbs* of herbaceous plants are frequently sought after by mammals, for the sake of the store of starch-food they contain. We can therefore understand the advanced defensive position taken up by such species as the Poison-bulb (*Buphane toxicaria*), an amaryllidaceous plant, in which this part is so poisonous as to be fatal to cattle. Many roots of plants are frequently poisonous.

The most abundant secretion of plants, especially of the larger arboreal kinds, is the formation of tannin, a substance peculiarly objectionable to animals of all kinds. Although usually found chiefly in the bark of trees, it is by no means limited to that portion. Its presence there can be easily understood, for if the bark were gnawed all round by ruminating animals (as they have a tendency to), the tree would be destroyed. It is between the protective bark, and the woody stem, that each year's layer of annular growth is formed. An extra secretion of *tannin* in the bark, therefore, must be highly protective; and how much is formed there is indicated by the fact that our commercial supplies are derived chiefly from the bark of trees alone.

Tannin, however, is found in its most concentrated state in certain fruits, such as the acorn, walnut, beechnut. It forms in the dark-brown skin, which all eaters of walnuts are careful to peel away before partaking of the white flesh; although they may not be aware that this part, defending like an inner wall after the hard shell has been broken through, is the counterpart of that other inner defence, known as the "stone" (*endocarp*), found in such pulpy fruits as the plum; whilst the hard shell, and, in the case of the acorn the tough and leathery outer skin, is the equivalent to the outer layer of succulent and juicy flesh, for whose sake the plum is cultivated.

Tannin is more or less abundant (and its presence must always be explained on the ground of its being a defence against various enemies), in the Oak, Elm, Willow, Elder, Plum, Cherry, Sycamore, Birch, Poplar, Hazel, Ash, etc., the three first possessing it perhaps in the highest degree. It abounds in the wiry but woody tissues of the Common Heath, from which it can be extracted by boiling. Some of the tropical *Acacias* contain even a larger proportion of it than our Oak bark.

Few plants have tannin more generally diffused throughout their tissues than Ferns. In the Bracken (*Pteris aquilina*),—a species which has been little altered since the Carboniferous Period,—it is very abundant, although most concentrated in the so-called root or *rhizome*. How small a proportion of mammals or caterpillars feed on the abundant green fronds of Ferns! Perhaps the secret of their freedom from such attacks is the highly defensive position in which they have been placed by their diffused secretion of tannin in every part. An English common, with its gorgeous investment of Heath and Bracken, is therefore a good illustration of the law of natural selection. Both those plants are rich in tannin, and both are comparatively untouched, although wild animals of various kinds abound there, glad to partake of such scanty food as they can find, and to utilise the vegetation they dare not eat, into a shelter and a home. Perhaps, also, one reason why Ferns have such a world-wide geographical distribution, and also why they have survived in practically an unmodified state, through all the various geological and biological changes our old planet has passed through, may be partly due to that very protection which so effectually screens them now from animal foes.

The secretion of *silica*, which we sometimes find in inordinate quantities in the leaves and stems of plants widely separated by natural affinities, is only understood when we know what an effectual protection it is from caterpillars and other depredators. Among familiar plants it is perhaps most abundant in the cuticles of the Horsetails (*Equisetaceæ*), and various species of grasses, sedges, etc. The "Dutch-Rush" (*Equisetum hyemale*) was formerly imported into this country for the purpose of polishing metals, which quality it owed entirely to the silica in its composition. The Equisetums are doubtless allied to the ancient Carboniferous *Calamites*, and the genus Equisetum itself has probably been in existence since the Oolitic Period. Possibly the species owe their preservation because they have learned to secrete silica; just as we have surmised ancient types of Ferns have been perpetuated through their possessing *tannin*.

Many species, including the Bamboo and other Grasses, Canes, Palms, etc., must be favourably advantaged by their ability to secrete a mineral substance so disliked by animals as silica is. Many of the leaves of our common British Grasses, such as the Tufty-hair Grass (*Aira cæspitosa*), abounding usually in damp meadows and shady woods, contain it so largely that no cattle eat it from choice. A still worse species is the better known "Couch-grass" (*Triticum repens*), pest to farmers, for its leaves are innutritious and produce scour in cattle; and the plant can propagate itself by the nodes of its insidious creeping stems as well as by seeds; so that cutting it up means helping it to multiply! The "Prairie-grass" now transferred to grace our shrubberies, with its drooping blades and handsome white floral panicles, contains so much silica that its leaves cut like a razor if drawn rapidly through the hand.

The needle-shaped leaves of Fir and Pine-trees, and of so many other Coniferæ, are also rich in silica, and both their shapes and secretions protect them from browsing mammals and the attacks of caterpillars. Perhaps also they are further protected thereby from the devastations of parasitic fungi, as few species are known on Coniferæ. It should be further remembered that the Coniferæ are the oldest, geologically speaking, of all woody trees; just as the Grasses are perhaps the

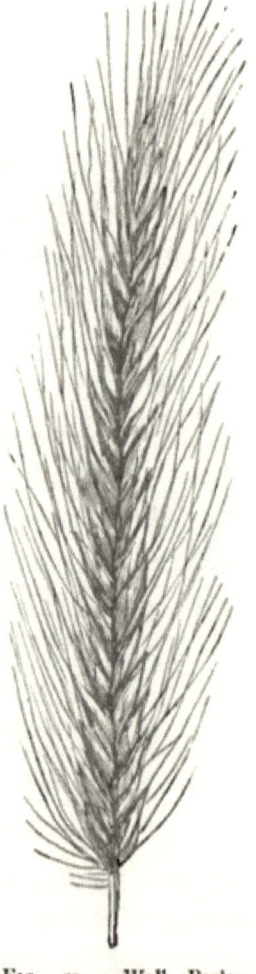

Fig. 51.—Wall Barley (*Hordeum murinum*), showing defensive "beard."

most ancient of true flowering plants. The wide distribution and geological antiquity of species of plants distinguished by the abundance of protective secretions should not be overlooked by the botanist.

It is noteworthy that, in nearly all the Grasses, the largest deposits of silica take place in the chaffy glumes which protect both the flowers and the seeds. These organs, the equivalents of the beautifully coloured floral segments of the Tulip and the Lily, have a practical and utilitarian duty to perform. Hence their absence of beauty and nutrition is in their favour. In some cereals these scales adhere quite close to the seed-corn, as in Oats; in others they terminate in formidable awns, charged with silica, as in the bearded Barley. In every instance they are protective to seeds known by all animals to be rich in nourishment. Occasionally these dry chaff scales, faithfully adherent to their charge even

FIG. 52.—Glumes and awns of Wall Barley.

when it has fallen to the ground, cork-screw it into the soil, on account of their sensitiveness to moisture, the seed with its scales slowly rotating on its axis, until it is thus placed out of sight and in a situation where it can sprout.

The true Rushes (*Juncus*) are leafless, but their green stems are uneaten because of their siliceous character. The same is true of the *Carices*, whose similar mineral protection is intensified by the *triangular* shapes of their stems, etc. The handsome blades of such species of water-reeds, as *Arundo phragmites*, are seldom attacked except by *mining* caterpillars, which have learned to penetrate between the siliceous cuticles.

The creeping habit adopted by many species of grasses and other herbaceous plants is a trick of self-defence. Under the Arctic snows a species of Pine always grows recumbently, so that the winter snows cover it up and protect it from Polar frosts. The nearly recumbent position assumed by the well-known Alpine Pine (*Pinus pumilio*), so abundant just beneath the snow-line in Switzerland, is only to be explained by reference to its advantageous condition in winter. Such a position must also be beneficial to it in summer, when it cannot be uprooted by violent winds.

One device adopted by fruits and seeds for protection against their numerous animal foes almost comes under the head of "Mimicry." Attention was drawn in a previous chapter to the fact that fruits were roughly divisible into two kinds, like flowers—viz. conspicuous and inconspicuous, according as the dissemination of their seeds depended upon organic or inorganic agency. The latter group is not only uncoloured, but actually evades notice. Whilst hanging on the tree or shrub which bears them such fruits are usually of the same green tint as the foliage, and are therefore concealed from view. In some cases, such as the Hazel (*Corylus avellana*), the fruits are enveloped in large green leaf-like bracts, which hide them from view by assimilating their appearance to that of leaves. But, as Grant Allen has shown, all of these fruits are of a dark brown or *ground colour* when ripe. As they lie where they drop on the ground, such a tint must prove more or less protective. Moreover, it is this class of fruits whose husks are generally either poisonous or uneatable. In many instances, particularly in tropical regions, where fruits have the most agile of all animals, the monkeys, to guard themselves against, the husk develops a great deal of lignine, and grows very hard, as in the Coco-nut. It will also proceed further (as in the fruit of the Brazil Nut, *Bertholettia excelsa*), and not only form a hard, almost impenetrable pericarp, but wrap each separate angular "nut" round with a coating of the same material as well. It is necessary to open a specimen of the fruit of the *Bertholettia* to fully understand the advanced degree of protection it has acquired against the assaults of monkeys. The "Cannon Ball-tree" (*Couroupita guianensis*) indicates its superior hardness by its very name. The Monkey-Pot tree (*Lecythis ollaris*), so delightfully described in Kingsley's *At Last*, is another fruit evidently specialised with reference to the quadrumana.

Most, if not all the fruits popularly known as "nuts," are included in this category. Our European species have to be protected more especially from squirrels, mice, and other small mammals. In addition, all are liable to the attacks of weevils.

As we stroll along some green lane-side, or across the breezy heath, in the glowing summer time, we cannot help noticing how many enemies the wild plants have to contend against. The leaves are gnawed by mammals, or eaten into tatters by innumerable caterpillars. By and by, when the summer heat is intensified, perhaps a drought sets in, and the poor patient leaves droop, become brown and shrivelled, and so drop off before their time. Meantime, how remarkable it is that the flowers on the same plants are neither eaten by animals nor much affected by heat? If we consider the great importance of the reproductive organs, it is evident that, under other circumstances, the species would soon be extinct. What is the nature of the subtle means of protection possessed by the petals of flowers, that they should be preserved amid such a wholesale vegetable massacre and plunder? Our curiosity is not abated on collecting a quantity of the most delicate, attractive, and soft petals of all kinds of flowers, and finding that hungry cattle utterly refuse to eat them, and that most caterpillars, voracious creatures, will actually die of hunger rather than partake of this kind of vegetable food! Evidently all conspicuous flowers have acquired at least *one* victorious method of completely checkmating the animal kingdom!

We actually employ the petals of certain kinds of flowers, such as the Flea-banes, as insecticides; and these compose a great part of some of the powders denominated "insect-killers" and "moth-banes."

Kerner, in his delightful little book, *Flowers and their Unbidden Guests*, explains the reason of this floral protection as follows: "The substances which make the flowers nauseous to many animals, and by which ruminants in particular are kept from them, are sometimes alkaloids, sometimes resins, but chiefly *ethereal oils*."

And yet these ethereal oils, whilst deterring caterpillars and grazing animals, *attract* flying insects from a distance by their perfumes! Kerner further states that "the chemical compounds which prevent many animals from touching fresh flowers are either volatilised, or undergo change, when the petals are dried. Many flowers, when dried, lose their special scent, or change it, and, mixed in the hay, are then eaten by ruminants without hesitation."

So far it is evident that the perfumes of flowers perform a double duty—that of informing insects at a distance of their neighbourhood, and that their flowers are open and the nectar ready; and, further, of protecting them from devouring caterpillars and browsing mammalia. It is possible they subserve a *third* end, that of screening the delicate petals from the consuming heat of the sun; for these structures remain untouched even when the hardier leaves are withered up and browned.

Six years ago I ventured an explanation of this phenomenon, in *Flowers, their Origin, Shapes, Perfumes, and Colours* (p. 311); and subsequent observation has only confirmed me in the opinion there expressed, and which may be here referred to in explanation. Professor Tyndall has shown the power which a spray of perfume possesses, when diffused through a room, to cool it; or, in other words, to *bar out* the passage of the heat rays. May not the possession by flowers of the ethereal oils or perfumes be the means of effectually protecting them from solar heat, and so enabling them to keep open unimpaired, until the purpose for which they expanded has been effected?

Every one knows that flowers possess different and characteristic perfumes. Lubbock, Bennett, and other botanists have proved the partiality of the more highly-cerebrated insects, such as bees, for certain *colours*—blue, for example. There is reason to believe that a difference in *odours* may be of quite as much importance to flowers, in guiding the proper kinds of insects to them, as difference in colours has proved to be. To night-opening flowers characteristic perfumes must be of supremest importance; for there is no colour visible in the dark, to attract or direct the attention of moths to them.

How odours can be specialised is illustrated by the *Stapelias* and some *Arums*, the former common in green-houses. The smell given off by the flowers of these plants resembles that of animal matter in a "high" condition. The lurid flesh colour of the flowers assists in the deceit; and it is no unfrequent thing to see them "blown" by such flesh-feeding flies as our common blue-bottles. Singularly enough, most of the flowers which give off fetid odours are either some tint of red or on their way to it; and such are generally visited by dipterous flies, even when only a faint odour is given off, if the flowers are of a liver colour and little known (on account of their recent introduction) to the insects of a place. For instance, last summer I was particularly struck with the fact that none but dipterous flies visited the male and female flowers of the variegated Laurel (*Aucuba japonica*). For several years past, in my own garden, the female plants have borne abundant crops of scarlet berries. On examining the flies taken from the male flowers, I invariably found their bodies scattered over with the pollen-grains. The liver colour of both male and female flowers had successfully attracted this class of insects, to whom this tint is associated with the animal food they like.

The trick adopted by the Lords and Ladies (*Arum maculatum*) of our hedgerows is worth remarking. If we peel off the neatly-wrapped spathe, we find inside, just at the part where it is constricted, and before it swells out again, a large number of hairs arranged so that insects can easily crawl in, but cannot get out, after the manner that eel-traps are constructed. The tall, *purple*-coloured spadix rises like a column in the midst, and the flies which have crawled into the trap are kept close prisoners for a day or two until the stamens clustered on the spadix have discharged their pollen, and it has fallen down to the bottom and dusted the poor prisoners! But "poor" it is hardly correct to call them, for they have been well treated in their confinement; somewhat after the manner voters were kept feasting at the candidates' expense, in the "good old times," when contested political elections lasted three weeks! When the pollen of the *Arum* has been spent, the hairs shrivel, and the imprisoned insects are set at liberty to carry the pollen to another plant, where perhaps the pistils are ready to receive it. How many generations of ancestors of the Common Arum must have passed away before this trick was brought to its present degree of perfection! Singularly enough, the rarer Birthwort (*Aristolochia clematitis*) has exactly the same kind of "eel-trap" arrangement as the Arum employed for the same purpose.

The numberless means which have been found out by plants to evade or checkmate their enemies is marvellous. Kerner's book is crowded with examples; but perhaps those which strike one most, as partaking of a certain degree of cunning, are such devices as that adopted by the common Goat's-beard (*Tragopogon pratensis*), which closes its flowers so regularly at noon-time that the ploughboy regulates his mid-day meal-time by it. That flowers open and close at different times of the day has been a well-known botanical fact since Linnæus amused himself by constructing his "Floral Clock." Country children are acquainted with the closing times of certain flowers, which they poetically term "going to sleep." Nearly ten years ago Sir John Lubbock drew attention to this habit, and expressed his opinion that it had some reference to the appearance of the insects which most benefited the flowers, as well as to their protection from others that were disadvantageous. Kerner's personal observations have since verified the acumen of the latter part of Sir John's observation in a very remarkable manner.

It is well known that ants are as fond of nectar as bees or butterflies, and they will get at it if possible, just as cats will to cream. But few or no creeping insects are advantageous to flowers, for the pollen will not adhere to their smooth bodies, even if they could travel faster and be more useful. Moreover, if ants get at the supply of honey they devour it, so that none is left for the flying insects for whom it was specially secreted. Hence flowers robbed by ants would stand a poor chance of being effectually cross-fertilised.

In other words, flowers have to deal with, and their structures are related to, not only the visits of "welcome guests," but to those of "unwelcome visitors" as well. Ants are among the most persistent of the latter class.

But these ants are *late risers;* they cannot stir abroad early because the dew is on the ground, and the leaves and stems, maybe, are dripping with it. Consequently the flowers of such plants as the Goat's-beard (*Tragopogon*) take advantage of the fact (being instinctively as well acquainted with this part of Formican Natural History as its modern chronicler, Sir John Lubbock), and so they open early and close at noon-day. The Nipplewort (*Lapsana communis*) and *Crepis pulchra*, abundant on all hedge-banks, where ants are usually numerous, open only from half-past five to nine or half-past—about four hours—in the morning; perhaps because they grow in drier spots.

But the most widely-adopted protective contrivance employed by plants to protect their flowers is the secretion of some sticky substance on the stems and calices of flowers, which acts as a kind of "lime," in which the greedy ants are sure to be caught and killed. The order *Caryophyllaceæ* has been the most successful in developing this device. Many species go by such popular names of "Catchflies," on account of numberless insects having been seen adhering to their sticky stems. Kerner gives a long list of viscid-stemmed and calyxed plants. How effectively this method defends them from entomological depredators is proved by his counting sixty-four small insects sticking to a single inflorescence of *Lychnis viscaria*. In the Gschnitz Valley, Tyrol, he tells us he collected over sixty *species* from the viscid flower-stems of *Silene nutans* alone, of which a large number were ants. Suffice it for us, however, that the reason for the secretion of these sticky fluids by plants is a defensive one; just as the secretions of poison, tannin, and bitter principles are now known to be.

I cannot forego mentioning a special example, indicative of what would be regarded as *sagacity* of the acutest kind if it had been exhibited by an animal. One of our commonest British plants is the Amphibious Persicaria (*Polygonum amphibium*), which grows, as its botanical specific name indicates, as well on the dry land as in water. All the specimens growing on land possess sticky glands, whose exudations protect the flowers from crawling insects. But when it grows in water no viscid secretion is elaborated—as if the plant were conscious that the water was effectual protection enough from all creeping insects!

Some plants prefer paying "black-mail" to such predatory enemies, so that they may be left to their lawful pursuits. They accordingly form nectaries on their leaves, which secrete sweet fluids. Several familiar plants, such as the Bean (*Vicia faba*), some Vetches (*Vicia sepium* and *Vicia sativum*), are illustrations of this endeavour to divert the enemy, or bind him over to keep the peace, by paying a special tribute. Kerner mentions, among other familiar species, *Impatiens*, *Ricinus*, *Viburnum tinus*, and *Viburnum opulus*, in all of which the leaves produce nectar for the entertainment of ants, etc. In some instances the nectar is produced by a special group of epidermal cells, transformed into glandular tissue, as in *Prunus Laurocerasus*; occasionally it is secreted by special *trichomes*, or hair-like organs, on the surface of the leaf or its stalk.

Belt (*The Naturalist in Nicaragua*) has given perhaps the most surprising instance yet known of how a plant has adroitly converted what are generally a nuisance and pest into an extra means of security, by affording food and lodgings to ants, instead of killing them or driving them away:—

"A species of *Acacia*, belonging to the section *Gummiferæ*, with bi-pinnate leaves, grows to a height of 15 or 20 feet. The branches and trunk are covered with strong curved spines, set in pairs, from which it receives the name of 'Bull's-horn Thorn;' they having a very strong resemblance to the horns of that quadruped. These thorns are hollow, and are tenanted by ants, that make a small hole for their entrance and exit near one end of the thorn, and also burrow through the partition that separates the two horns, so that the one entrance serves for both. Here they rear their young, and in the wet season every one of the thorns is tenanted; and hundreds of ants are to be seen running about, especially over the young leaves. If one of these be touched, or a branch shaken, the little ants swarm out from the hollow thorns and attack the aggressor with jaws and sting. They sting severely, raising a little white lump that does not disappear in less than twenty-four hours.

"These ants form a most efficient standing army for the plant, which prevents not only mammalia from browsing on the leaves, but delivers it from the attacks of a much more dangerous enemy—the leaf-cutting ants. For these services the ants are not only securely housed by the plant, but are provided with a bountiful supply of food; and, to secure their attendance at the right time and place, this food is so arranged and distributed as to effect that object with wonderful perfection. The leaves are bi-pinnate. At the base of each pair of leaflets, on the mid-rib, is a crater-formed gland which, when the leaves are young, secretes a honey-like liquid. Of this the ants are very fond, and they are constantly running about from one gland to another, to sip up the honey as it is secreted. But this is not all: there is a still more wonderful provision of more solid food. At the end of each of the small divisions of the compound leaflet there is, when the leaf first unfolds, a little yellow fruit-like body, united by a point at its base to the end of the pinnacle. Examined through a microscope, this little appendage looks like a golden pear. When the leaf first unfolds the little pears are not quite ripe, and the ants are continually going from one to another, examining them. When an ant finds one sufficiently advanced it bites the small point of attachment; then, bending down the fruit-like body, it breaks it off and bears it away in triumph to the nest. All the fruit-like bodies do not ripen at once, but successively, so that the ants are kept about the young leaf for some time after it unfolds. Thus the young leaf is always guarded by the ants, and no caterpillar or larger animal could attempt to injure them without being attacked by the little warriors. The fruit-like bodies are about one-twelfth of an inch long, and are about one-third the size of the ants, so that the ant bearing away one is as heavily laden as a man bearing a bunch of plantains. I think these facts show that the ants are really kept by the *Acacia* as a standing army to protect its leaves from the attacks of herbivorous mammals and insects."

In other words, this Acacia can afford to provide food and lodging for an army of insects employed to defend it by stinging, instead of elaborating stings of its own and using them, as our common species of Nettles do. Belt mentions several other kinds of plants he came across in Central America, such as a species of *Melastoma*, of *Passiflora*, etc., which secrete nectar and provide shelter for ants to defend them against the dangerous leaf-cutting species.

The most elaborate defences against the depredations of ants and other "unwelcome guests" are to be found, however, in the crowds of delicate white or coloured hairs, usually arranged in crowded "weels" inside the throats of flowers. Sometimes they spring from the inner surfaces of the tubed corollas, occasionally they form a *chevaux de frise* on the base of each stamen; thus producing a very complex interlacement, which adds much to the

FIG. 53.—Bog-bean (*Menyanthes trifoliata*).

beauty of the flower, and proves a great barrier against small creeping insects getting at the honey. In the Bog-bean (*Menyanthes trifoliata*), the upper surfaces of the pink petals are fimbriated all over, to prevent unwelcome insects crawling over them. In the Gentians the inner arrangements of protective hairs cannot fail to strike a young botanist with

Fig. 54.—*Filago canescens* (Jord).   Fig. 55.—*Filago spathulata* (Presl).
Both these species are covered with dense short hairs.

admiration at their exquisite beauty. Few flowers habitually fertilised by flying insects, are without these "inner guards" to their nectaries.

Not unfrequently both leaves and stems are *tomentose*, or woolly. This device proves a splendid defence against *Aphides*, or plant-lice, whose immense power of reproduction and suctorial appetites render them among the most persistently annoying enemies the higher vegetation has to cope with. Smooth-stemmed plants have no chance with them; and it is noticeable that the introduced and cultivated plants in our gardens are more liable to their attacks than wild species. We do not find the wild Roses

Fig. 56.—Herb-Robert (*Geranium Robertianum*). Stems covered with hairs.

of our lanes so thickly encrusted with strata of plant-lice like those commonly observed on our garden Roses.

The crowded hairs on the stems of plants, as well as the viscid secretions already mentioned, are of great advantage to the species possessing them.

These hairs vary in their character and density until we find them passing into "prickles." Some species have converted the hairs into "stings," as witness the Nettles. No better illustration of the thorough triumph of this contrivance could be afforded than the abundance of Nettles by our waysides where every other kind of vegetation has been cropped and eaten by passing animals; but the Nettles are left alone, except by caterpillars. Not a few British plants have learned the value of a Nettle neighbourhood, and are always to be found growing near or even with them; just as the helpless peasants grouped themselves under the shelter and protection of the barons' castles in the Middle Ages.

In addition to those numerous and greedy enemies of plants, the mammalia, vegetation has to be protected from another class of depredators—slugs and snails. How voracious they are gardeners well know, and accordingly set traps innumerable to catch them, or arrange deterrents to keep them off. In a state of nature plants adopt the same device— developing poisons in their tissues, or becoming unsuitable to their molluscan appetites. A better protection still is the formation of *thorns* and *prickles*, over which slugs and snails find it impossible to trail their soft bodies without injury. In some plants—the Brambles, for instance—these prickles assist in climbing; in others they keep off browsing mammals as well as snails.

These "prickles" and "thorns" have been evolved

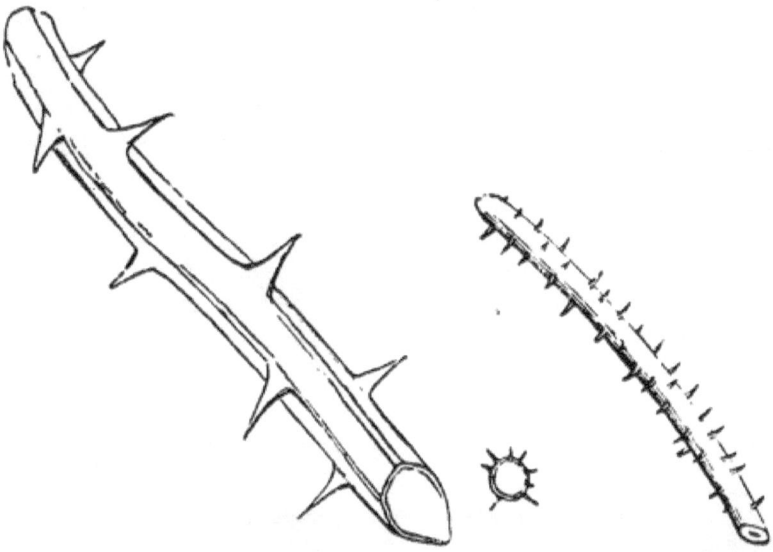

Fig. 57.—Portion of stem and transverse section of *Rubus rhamnifolius*.

Fig. 58.—Portion of stem and transverse section of *R. cæsius*.

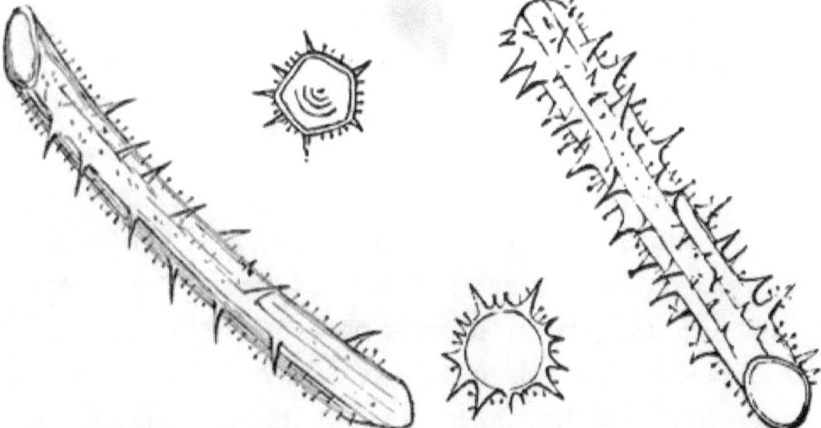

Fig. 59.—Portion of stem and transverse section of *R. glandulosus*.

Fig. 60.—Portion of stem and transverse section of *R. rudis*.

from many sides. Sometimes they are merely

stiffened hairs, as in the Gooseberry and Bramble. Or they are produced by aborted branches, as in the Sloe and Whitethorn; and at other times they are the result of the complex changes which *leaves* have undergone. The latter are perhaps best shown

   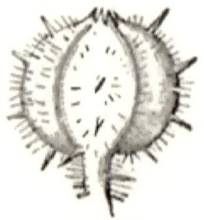

FIG. 61.—Flower-bud of *R. nemorosus*.   FIG. 62.—Sepals of *R. cæsius*.   FIG. 63.—Flower-bud of *R. fusco-ater*.   FIG. 64.—Flower-bud of *R. umbrosus*.

All these flower-buds possess defensive prickles and stiffened hairs.

in the *Cacti*, where the stiff, needle-like thorns in reality are aborted leaves, whose duty is now purely defensive, and whose carbon-absorbing work is deputed to the succulent green stem, provided with stomata for the purpose.

At the Cape Euphorbias assume quite a Cactus-like shape. One of them (*Euphorbia grandidens?*) is described by Sir Charles Bunbury (*Botanical Fragments*) as follows: "When seen at a great distance its general form is not very unlike that of a Pine-tree, though it is extremely different on a near view. It has no leaves, but its young branches are extremely succulent, green, thick, angular, like those of a *Cactus*, and beset along the angles with pairs of short spines. . . . Its flowers, which are of a greenish-yellow colour, comparatively small and inconspicuous, are seated on the angles of the branches. The whole plant is full of an excessively acrid and caustic milk, which gushes out in great quantities whenever an incision is made."

One of the most remarkable features about the arborescent Cape flora is the degree with which it is defended by thorns, etc. When we remember the extraordinary numbers of wild herbivorous animals which formerly swarmed in this part of the world, and how in seasons of drought they would be ready to eat anything, we see the use to the plants of such defences. An extremely common shrub is *Cliffortia ruscifolia*, whose prickle-pointed leaves are as sharp as needles, and break off at the points, so as to remain sticking in the flesh.

The best illustration of the conversion of leaves into protective spines in this country is afforded by the Gorse or Furze (*Ulex europæus*), abounding on all our commons, in spite of nibbling rabbits and crowding sheep. The early development of the leaves of the Gorse is *ternate*—that is, they are composed of three leaflets, like other leguminose plants; and the process of the gradual stiffening of the subsequently-formed leaves into protective spines can easily be studied by any one who will take the trouble to sow a few Gorse seeds and watch their development. No better proof could be desired of the gradual transformation of a leguminose plant into this defensive condition. The Broom (*Sarothamnus scoparius*) has proceeded some distance in the same direction, but has not yet succeeded in effectually stiffening its leaves. Perhaps this is not required, as the leaves secrete an objectionable substance instead. The Barberry commences life on quite contrary principles to the Gorse, for its first pair of leaves become *spines*, from the axils of which the next pair of true leaves shoot.

The rank growth of the Gorse on every common, haunted though it may be by rodents and ruminants, proclaims the success which has attended its leaf-transformation. But *absolute* success is not known in this world, and the wily Gorse is no exception to the general rule. A plant of which a more extended notice will be taken hereafter, is to the Gorse more than what Mordecai was to Haman—it has completely baffled and defeated it. I allude to the Dodder.

Many plants possess leaves whose edges curl and stiffen into defensive spines, so that few or no browsing animals will partake of them, especially if any other vegetable food is at hand. The Thistles, Eryngo, Teasels, etc., have adopted this plan to perfection—the Holly to a fairly successful degree. For the leaves of the latter are soft and succulent at first, and therefore easily fed upon at that time. But the Holly leaves happen to be soft when other and more nutritious food is abundant, and so the attention of browsing animals is called off them. When other vegetable supplies have failed, and the Holly, being an evergreen, attracts attention by its glossy green leaves, the latter have been stiffened and protected by the well-known prickles in the interval! Southey's familiar lines, therefore, are only partly true, although they are remarkable in showing that he fully recognised the reason for the defensive character of Holly leaves :—

> "Below, a circling fence, its leaves are seen,
>   Wrinkled and keen ;
> No grazing cattle through their prickly round
> Can reach to wound ;
> But as they grow where nothing is to fear,
> Smooth and unarm'd the pointless leaves appear."

Belt graphically describes a similar defensive kind of vegetation existing in Central America. "In the midst of these plains grow spiny Cactuses, low leathery-leaved trees, slender, spiny Palms, prickly Acacias, and thorny Bromelias. This spiny character of vegetation seems to be characteristic of dry rocky places, and tracts of country liable to great drought. Probably it is as a protection from herbivorous animals, to prevent them browsing upon the twigs and small branches when vegetation is dried up." Grant Allen, in commenting on the prickly character of the leaves of our common Saltwort (*Salsola kali*), in *Vignettes from Nature*, goes fully into the matter. "Sand-loving plants are naturally exposed to very great danger from herbivorous animals, against which they are accordingly compelled to protect themselves by some hostile device. In the first place, there is comparatively little vegetation on sandy spots, so that each plant runs an exceptional chance of being eaten. Then, again, the succulence and juiciness of sand-haunting weeds makes them particularly tempting to thirsty animals, which are sure to eat all unprotected specimens. Hence, as a rule, only those survive which happen to have developed some unpleasant personal peculiarities. Many sand-haunting or desert plants are more or less pungent, or have disagreeable alkaline essences stored up in their leaves ; and these alkaline constituents, which they easily obtain from the soil, formerly caused many of them, Saltwort and Glasswort among the number, to be burnt for barilla. But most sand-loving weeds have solved the difficulty in another way by simply acquiring thorns or prickles. In the Saltwort each leaf ends in a stout spine, which of course runs into the nose of any too inquiring cow or donkey. In the West Indies, Cactus hedges line all the roads in the plains, and rise in a solid wall to the height of 15 or 20 feet. No animal on earth dare attempt to pass through such a hedge, and the task of cutting one down, when necessary, is extremely difficult. On bare dry expanses, like the Mexican plain, Cactuses and Agaves run wild in every direction, collecting what little moisture they can in their thick stems or big succulent leaves, and defending it against all herbivorous enemies by their formidable spines. To prevent evaporation they are covered with a thick and very firm epidermis, so that they lose very little of their moisture, even during months of drought.

"What these great desert plants do on a large scale, our little English Saltwort does on a much smaller scale. It has the same strong prickles, the same thick, juicy leaves, the same protective epidermis, and the same general aspect of growth as the Cactuses themselves. If one were to enlarge it twenty-fold, every casual observer would set it down as a desert species at once."

Perhaps one reason for the succulence of such maritime plants as the Saltwort is the quantity of salt entering into their composition. This substance is remarkable for its power of absorbing moisture, even from the atmosphere, and it must, therefore, be a gain to succulent plants to grow where salt abounds. In proof of this we have the fact that well-known maritime plants have settled down near saltworks inland, as in Cheshire and Worcestershire. Moreover, maritime plants are notable for their fleshiness and succulency, although many widely different orders are represented by them.

Various other peculiarities distinguish plants in widely-separated orders, all of which have been developed in the keen and seemingly cruel "struggle for existence;" without such a "struggle," however, the organic world would never have attained a much higher rank than that of scarcely differentiated particles of protoplasm. As observation is extended, every characteristic of plants will be found to bear reference to its present or its past wellbeing and protection. Some of the characters are still retained in their structures, although no longer necessary or requisite, just as we keep up certain fictions in law and etiquette — "survivalisms" of once important functions.

The trifoliate leaves of the Clovers, Oxalis, and a few other plants, close at night; and there can be little doubt that by doing so they present a less surface to the late frosts which often linger on to the season when the leaves are green and in active work. The leaves of many Australian trees have acquired the trick of turning the edges of their leaves upwards, for the contrary purpose of exposing as little surface as they can to the dry and intense heat of the sun. To this end their leaf-stalks give a half-twist, and so continue to present the edge only to the sun as it travels across the heavens. The *stomata* or carbon-feeding mouths of their leaves, instead of being placed on one side (generally the lower) of the leaf, in the leaves of these Australian shadeless trees, are arranged on both sides. By thus exposing a limited, indeed a minimum, surface of leaf to the sun, such trees prevent the evaporation of their moisture—a most important incident, when we consider the long droughts occurring in such countries.

Even the milky juices exuded by the leaves of the common Lettuce, Milk-thistle, etc., may be frequently ranked among the defensive arrangements of plants. Kerner narrates some experiments he made which prove that ants, and other insects crawling over such leaves, soon get glued down by the milky exudations produced by the claws of their tiny feet. His experiments were made with *Lactuca augustana* and *Lactuca sativa*. He says: "No sooner had the ants reached the uppermost leaves, or the peduncles and the involucral bracts, than at each moment the terminal hooks of their feet cut through the epiderm, and from the little clefts thus made milky juice immediately began to flow. Not only the feet of the ants, but the hinder parts of their bodies were soon bedrabbled with the white fluid; and if the ants, as was frequently the case, bit into the tissue of the epiderm in self-defence, their organs of mastication also at once became coated over with the milky juice. By this the ants were much impeded in their movements, and in order to rid themselves of the annoyance to which they were subjected, drew their feet through their mouths, and tried also to clear the hinder part of their body from the juice with which it was smeared. The movements, however, which accompanied these efforts simply resulted in the production of new fissures in the epiderm, and fresh discharges of milky juice, so that the position of the ants became each moment worse and worse. Many of them now tried to escape by getting, as best they might, to the edge of the leaf and letting themselves fall from thence to the ground. Some succeeded, but others tried this method of escape too late; for the air soon hardened the milky juice into a tough brown substance; and after this all the strugglings of the ants to free themselves from the viscid matter were in vain. Their movements became gradually fewer and weaker, until finally they ceased altogether, and the dead animals were left adhering to the involucre or the uppermost leaves."

The milky juices of the Poppy (*Papaver somniferum*), which soon hardens into opium when the capsule is artificially scratched, is an illustration of the origin and natural function of similar defensive secretions, not necessarily provided against ants, with which other plants have gradually and successfully armed themselves.

CHAPTER IX.

CO-OPERATION.

IT was not in altogether a figurative sense that we regarded leaves as vegetable units, the equivalents of *zooids* in such polypidoms as a Sea-fir. Every plant, shrub, or tree may be looked upon as a colony of such units; a co-operative society on a varying scale, ranging from individuals possessed of merely a thallus, like Seaweeds, Liverworts, etc., to herbaceous plants with not more than half a dozen real leaves, and upwards to the lordly Oak and gigantic Euphorbias with their myriads. Vegetable society, like human, is therefore collected into villages, small towns, and populous cities. The leaves of a plant share in the mutual benefit of its life. A tree is a nation, with its units of leaf-population coming and going year after year. It is subject to the same laws of rise, decline, and ultimate fall, which history shows has characterised the national life of many countries. Its defensive thorns, prickles, spines, poisons, etc., are to it what such defensive resources as the army, navy, and fortifications are to a great nation. Its roots and leaves are the *trading* or wealth-accumulating members, its flowers are its expending members, its fruits the emigrating part of the population, its thorns and spines leaf-units set apart, as we do our soldiers, for the defence of the general community.

Weak animals find security in associating together, and mankind have doubtless had their own social character evolved, with all the qualities and attributes belonging to it, from adopting a similar habit. Herbivorous animals collect in herds, birds in flocks, and even insects often in dense crowds. The law which has produced such a tendency has operated in a similar manner among certain kinds of vegetation. Everybody has noticed how the Alpine Gentians, Anemones, etc., have congregated in brilliant patches in Switzerland, to the general exclusion of other plants. The Daisies and Buttercups of our English meadows, and the Poppies of our cornfields, flourish best when congregated in numbers.

A real and most important reason for such a social habit, is probably to tempt insects to their neighbourhood by exhibiting great masses of colour. Many of the "social plants," as Humboldt long ago termed them, have got possession of the ground through being specialised to extreme physical conditions, as the Arctic and Alpine flora are to cold, for instance. Height above the sea-level is equivalent in climate to high latitude. In our lowlands *seasonal* appearance is equivalent to both these conditions, and the earliest plants to blossom in this country belong to genera, and sometimes even species, which are Alpine, as *Chrysosplenium oppositifolium* and *C. alternifolium;* and thus we have a "vernal" flora as well as one peculiar to the summer. All of our spring plants that are identical with Alpine or Arctic species are distinguished by flowering *earlier* in the year; in some cases as much as two or three months before their brethren within the polar circle, or on the fringe of Alpine snowfields. Consequently, we owe our beautiful and cheering " spring flowers " to the same great physical geographical distribution which brought over the Arctic plants now found on our British mountain-tops. And they are as much acclimatised and protected by simply altering their flowering time, and blossoming earlier in the year, as if they had been left stranded on high elevations instead.

One chief characteristic of this class of plants is their power to endure extreme cold and wet. Several kinds of Buttercups, Stitchwort, etc., flower all the winter through, unless the season is unusually cold.

Many of these spring flowers display a tendency towards a seasonal "division of labour." They usually either flower before they leaf, as with the

Coltsfoot (*Tussilago farfara*), Crocus, and Snowdrops; or try to do so, as with the Primroses, Violets, and Daffodils, which develop their leaves much more abundantly after flowering than before. The Meadow Saffron (*Colchicum autumnale*), is a plant which produces leaves at one part of the year, and flowers at another.

Our "social plants" have been largely influenced in their gregarious habits by mutual likes and dislikes. What species extend over so large an area as the Heathers? Our moorlands and hill-tops are purpled with them in the late summer time, but they refuse to take up their abodes on limestone and chalk hills; and they show just as much preference for the Millstone Grit formation as they do dislike for all kinds of calcareous rocks. Sir Charles Bunbury (*Botanical Fragments*) tells us that "some of the Cape Heaths, like those of our own country, are social or gregarious plants, growing crowded together in large masses, and covering considerable spaces of ground." The well-known Cornish Heath (*Erica vagans*) extends only over the Serpentine rocks, whose area it designates like a geological map.

The plants usually growing profusely together near the sea, designated "social," may do so because of their salt-loving habits. Among these may be mentioned the Thrift (*Statice armeria*), the Sea Lavender (*Statice limonium*), the Sea Convolvulus (*Convolvulus soldanella*), Atriplex, Sea Buckthorn, etc. The Yellow

Gorse (*Ulex europæus*) takes possession of our sandy commons and heaths unasked, to share them here and there with the Heather. The Broom (*Sarothamnus scoparius*) has put in a claim to the embankments and cuttings of our railways; the Ragwort (*Senecio jacobea*) has long ago set its sign and seal on all badly-kept pastures, and its brilliant gold makes them rich in colour, if they are poor in produce. Thistles in abundance cover our waste ground, in company with Nettles and several other vegetable vagabonds, all of which are found in each other's company, because they love to follow the same habits of life. The "social plants," however, are much more numerous in temperate regions than in tropical. Kingsley says they are rare in tropical forests, the only instances he saw being the Moras and the Moriche Palms. In this respect the Cape of Good Hope seems to be an exception, for the marvellous flora of the "bush" is greatly of a social character. Sir Charles Bunbury states that a large *Mesembryanthemum*, with bright green leaves and large straw-coloured flowers, is one of the commonest of all plants on all the sandy lands. Its stems, lying flat on the ground, spread so as to form extensive mats of verdure more lively than that of the surrounding vegetation.

We have already seen that botanists consider all the parts of flowers as so many modified leaves—a theory which most of the "monstrosities" and "sports" of our garden plants corroborate. It appears to be a law that whenever large numbers of living objects, whether of animals or plants, congregate together, some of them are modified for the benefit of the rest. The transformation of leaves into floral organs, spines, tendrils, etc., is an illustration of the degree to which this is carried out in flowering plants. And, as we shall presently see, even when flowers themselves are grouped in large numbers, some of them undergo marked changes from their brethren. The fact distinguishing human civilisation, that the more advanced it becomes the more it is differentiated, is true also of plants and their flowers.

The main end and aim of all attractive flowers is, as has already been seen, for the purpose of inducing insects to visit them. The colours and shapes of flowers have been evolved for this purpose; but we do not know all the numerous reasons why they should have assumed so many different sizes.

Thus much, however, is evident, that flowers are usually gainers by being grouped together. It has been suggested that the reason why certain kinds of plants are "social" is that their combined masses of colour all the more certainly attract the butterflies which habitually cross them. This must certainly be the case with the Alpine Gentians, Anemones, and Violets, for bees do not fly to such altitudes. Moreover, it is probable that butterflies have not such keen sight as bees; and if so, we should expect that the flowers dependent on the visits of the former would gradually adopt a "social" or "gregarious" habit of life.

There is another sense in which flowers may be regarded as "social" organisms, and that in a much higher and more specialised form than the accident of mere congregation, viz.—when they cluster together on the same flower-stalk. All the methods

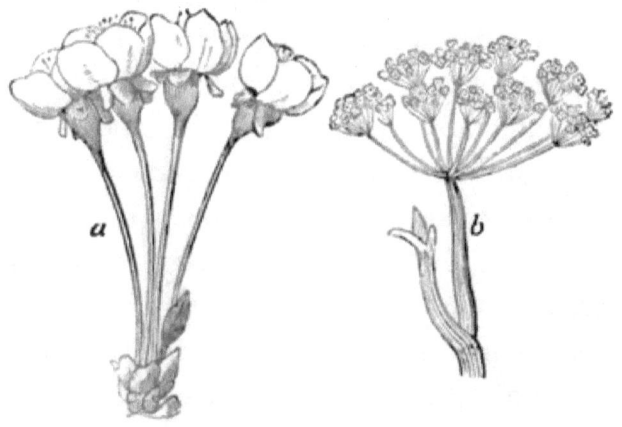

FIG. 65.—*a*, Simple umbel of Cherry; *b*, Compound umbel of Fool's Parsley.

of inflorescence known to and described by botanists, —" spike," "raceme," "panicle," "corymb," "cyme," "umbel," "capitulum," etc. — are only so many methods by which flowers are grouped together for mutual advantage. The numbers collected in these well-known arrangements vary exceedingly; but their number is uniformly found to be related to the size of the individual flowers. When the latter are large, the number of flowers clustered together will be few—when they are small they are generally numerous in proportion to their smallness.

In no order is this better seen than in the *Leguminosæ*. Compare the solitary flowers of the lovely Grass pea (*Lathyrus nissolia*) with the minute but similarly constructed flowers collected to form the "heads" of the Clovers and the Trefoils. No flowers are perhaps more specialised to the visits of the most intelligent of insects than those of Clover, but what would they be if they grew singly? Co-operation has been the secret of their success; as indeed it is of innumerable species in other orders of plants where the same plan has been adopted.

The *Umbelliferæ, Dipsaceæ*, and *Compositæ* have carried out this idea with the completest success, and with certain modifications of a most suggestive character; some of the members of their floral colonies being altered for the benefit of the community. This alteration has been carried to the extreme point of even sacrificing the individualities of some members for the wellbeing of the rest! Such an arrangement is by no means limited to the above-mentioned orders, even in our British flora. Thus that of the Honeysuckles (*Caprifoliaceæ*) has a considerable latitude of variation. The large flowers of the common Honeysuckle are usually grouped to the number of about half a dozen on the flower-head; whilst the smaller individuals in the Elder (*Sambucus nigra*), Wayfaring Tree (*Viburnum lantana*), and

Guelder Rose (*Viburnum opulus*), are clustered in great numbers. The grouping of the latter species is remarkable for the *increased size* of the outer circle of flowers—a normal condition, to which every member of the entire collection attains when the Guelder Rose is grown in our gardens, where its "snowballs" of flowers are prominent and pretty objects. But in the natural state all these large outer flowers are *barren*. Their size has been increased for the benefit of their brethren, so as to render them more conspicuous to insects, but they have sacrificed their own fecundity. Floral *altruism* is a fact in the vegetable kingdom, only found in the most differentiated floral societies; just as we meet with it only in the highest-developed of humanity, although we anticipate it will be still more developed as mankind grows out of its lower into its higher life!

This principle (for I cannot call it by any other name) is carried out in the highest degree among the *Compositæ*, and especially in that division of the order called *tubulifloræ*, of which the Daisy, Dahlia, Chrysanthemum, Chamomile, Sunflower, etc., are familiar examples. It need hardly be stated now that these objects are not *single* flowers—but colonies or collections of small flowers, all arranged on a disk or head; as may be best observed in the Sunflowers of our gardens, where both the florets and the disk attain their largest sizes. Hundreds of perfect little flowers, all of high organisation and development (for they are in a very advanced stage, and have had all their petals fused together into *gamopetalous* corollas), are thus collected together. Outside them are the "ray" florets, strap-shaped, and often, as in the Asters and Chrysanthemums, attractively and even gorgeously coloured; and sometimes, as in the Sunflower and Dahlias, growing to a great size. But they are all barren! Frequently they retain their sexual organs, but they are always aborted. In fact, these ray-florets live for the benefit, not of themselves, but of their more perfectly formed brethren. And in such floral colonies as our common Daisy the pink-tipped white "rays" not only attract flying insects to the minute yellow florets which compose the disk, but bend over and cover them up like nurses at night; as if for fear they should suffer frost, or be spoiled by rain or dew—and then the country children say the "Daisy has gone to sleep"!

FIG. 66.—Ox-eye Daisy.

The *Capitulum* of one of these plants is analogous to a beehive or ant-colony among insects; the latter are the highest cerebrated of all insects, just as composite flowers are acknowledged to be the most

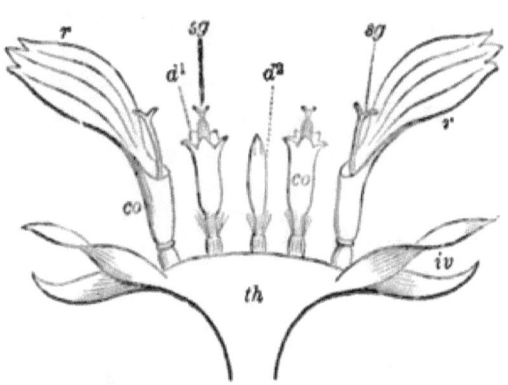

Fig. 67.—Section of Daisy; *r*, ray-florets (barren); *co*, tubular florets; *th*, receptacle; *iv*, segments of involucre.

wonderfully organised among plants. Their smallness has conduced to their social habits, as may have been the case with all kinds of inflorescences. The necessity for banding together has eventually evolved *altruism*, or the requisition for some members to live for the benefit of others. In a beehive or an ant's nest the sexless workers are the analogues of the barren "ray"-florets of the Daisy and the Sunflower.

The Knapweeds (*Centaurea nigra, C. cyanus*, etc.) have not advanced so far on the path of specialisation of the external florets as the Daisy and Chrysanthemum, whilst the Dandelion and Chicory (representatives of the *liguliforal* division of Compositæ) have proceeded a stage farther. The outer florets of the Knapweeds are still tubular, exactly like their fertile brethren, except being larger, and *barren*. Occasionally we get specimens of wild field Daisies possessing white *tubed* ray-florets—a reversion to an ancestral condition. And always, when Daisies are planted in rich soils, and kept clear of competition, as in our gardens, they become *double*— that is, the tubed florets of their disk are converted into strap-shaped ones, like those of the "ray"—and so partly assume the normal condition they have attained in the Dandelion, Hawkweeds, Chicory, etc.

The "umbels," as the floral colonies of the *Umbelliferæ* are called, approach the specialisation of the *Compositæ*, but not in the degree in which *altruism* is carried out; but most if not all of the outer flowers of this order have larger petals than the rest. So far, therefore, it would appear as if their character were *egoistic* rather than *altruistic*—that they were aristocrats, or plutocrats, rather better off than their neighbours. But at any rate the floral society of which they are members are gainers thereby, for insects are attracted to them all the more certainly for the extra conspicuousness. None of these extra-developed members

FIG. 68.—*Centaurea cyanus*, or Corn Bluebottle; *a*, enlarged barren florets.

of the community lose their fertility because of their increased size and prominency.

The *Dipsaceæ* are separated from the large and

FIG. 69.—*Angelica sylvestris*—showing compound umbels of small flowers.

cosmopolitan order of composite plants chiefly by the little incident that the pollen-bags, or *anthers*, crowded within the little tubed florets, do not cohere as in the *Compositæ*. Among some of the few members of this community, however, we find the same tendency as in the *Umbelliferæ* above mentioned. In the Scabious, and more especially in the Knautia (*Scabiosa succisa* and *Knautia arvensis*), the flower-heads are merely collections of individual flowers. In the Knautia it will be observed how the outer members are larger and more conspicuous; and their lower parts are expanded so that insects may alight upon them with the greater ease. The community, however, is advantaged by its outer members being more highly endowed and specialised. Nobody with common sense doubts this also is the case in human societies!

Notice has already been taken (page 82) of the reasons for the two different colours in members of the *Boraginaceæ*—the old and the new being relatively blue and pink. There we have an illustration of how the old flowers continue to help the juveniles even when their own purpose has been effected. Perhaps it may be hereafter found that the two methods of inflorescence known as "regular" and "irregular" sometimes assist in the same patriotic work. Nobody can doubt it is so in a cluster of Apple-blossom, for instance. The central flower opens first, and all round it is a ring of brilliantly coloured, unopened flower-buds. Observe how all the colour of the Apple-blossom is distributed over the *under* surface of the petal, whereas the *upper* surface is the place usually the most brilliant, as in the Buttercups, Speedwells, and all others. But in a cluster of Apple-blossom the unopened flowers, owing to the accident of their petals being coloured in the *wrong* place, enormously assist in the successful fertilisation of the central flower, and thus the *young* flowers perform the same helpful function which in the *Boraginaceæ* (Forget-me-not, Viper's Bugloss, Borage, etc.) is performed by the *older* ones. It cannot fail to be noticed that this central Apple-blossom is often the only one which bears an apple. All the rest "take their chance," so that every cluster of such blossoms preaches the precious doctrine of *altruism.*

I have said that leaves are "vegetable units." Flowers and their parts are only modifications of them. Hence, every floral organ is also a unit, specialised for a distinct work, whether that work be successful or not. A leaf-bud is as much a "colony" of leaves, in an inchoate and undeveloped condition, however, as the flower-head of a daisy is a floral community ; and the same *altruistic* law is observable in an unopened leaf-bud as in the *capitulum* of a composite plant. The most perfect of all vegetable organisations are those in which some members bear the burden of the rest, and I know nothing which more forcibly impresses one with the importance of this law than the leaf-buds of our commonest trees and shrubs.

If we carefully examined the bases where the leaves are attached, about the end of August, we should see the leaf-buds already forming, or formed. Of course the abundance of green leaves conceals them, or draws off our attention from them. But there they are, ready formed. We have only to carefully dissect one of the numerous winter leaf-buds to find that all the leaves which will develop the following summer are already present. All that the summer's sun-light and sun-heat will do will be to increase their size. Each little leaf, or pair of leaves, is snuggly wrapped one within another, in the most approved style of packing.

Examine some of the leaf-buds we may find upon any tree, and which in some trees, such as the Horse-chestnut, attain a great size. Before the examination is concluded, even unscientific observers will have been struck with something they had not observed before. Open one; we see the leaves, very dwarfed and undeveloped, but still all are there. How very helpless and weak they look! In the midst of winter they are of course in a feebler condition than during the summer when there are no frosts nor biting winds to nip them. How are such delicate and undeveloped leaves preserved? Any Horse-chestnut bud will furnish an explanation. Its outside parts are of a dark-brown colour, and very sticky. They completely cover in and protect the inner and very delicate undeveloped leaves.

These external wrappers are called "bracts." They are undeveloped leaves, or in other words, they are leaves whose duty it has been to sacrifice themselves for the benefit of those inner leaves they are thus protecting from the winter's cold. They will never see the "promised land" of next summer—will never wave green in the gentle summer breezes, or be visited by the singing birds. They will die in view of the "promised land," for when the increasing heat and light of the early summer cause the leaf-buds to develop and increase in size, these brown sticky bracts will be forced to drop off, and perchance we may see the ground underneath the trees strewn with them in April and May. The principle of *altruism* ("self-sacrifice" and "heroism") is thus abundantly represented even in the vegetable kingdom. The manner in which the bracts cover over and protect the winter leaf-buds is not confined to one method. Thus those of the Lilac are green, but tough and leathery; those of the Mountain Ash are lined with wool—a capital non-conductor of external cold as well as heat. In the larger leaf-buds we always find some contrivance of this kind to keep the extreme cold from affecting the delicate inner leaves. In very small leaf-buds, such as those of the Hawthorn, the bracts or external leaf-bud wrappers are of a reddish-brown colour, and when leaf-buds are so small there is no doubt their diminutive size is in their favour. They are not then so liable to be nipped by frost, nor even to be eaten by birds.

Co-operation, so far as it means differentiation for purposes specially beneficial to the community, is even further carried out among leaves than flowers. We have seen that they are frequently modified, not only into bracts, but into such other altruistic states as thorns, tendrils, etc., all of which exist for the wellbeing or defence, not of the individual, but of the plant or community.

The co-operation of the *perfumes* of small flowers, the odour of each of which would be scarcely distinguished by itself, must be of great advantage to them in attracting insects. It will be observed what a large number of species of plants bearing very *small* flowers have no other colour than white or yellow. It is almost certain that few others are visited by moths, for no other could be seen, and the larger proportion of white flowers producing sweet perfumes over those of any other colour, has already been commented upon.

The law of co-operation is often adopted by fruits. What else can we denominate the clustering together of the almost microscopical fruits of the Strawberry upon their succulent receptacle—of the association of such perfectly-developed "bird-fruits" as the Dewberry, Blackberry, Raspberry, Cloudberry, and others in a common cluster? Small fruits have assumed exactly the same associated and combined character as the small flowers of the Umbelliferæ and the Compositæ in the floral world. " Union is strength " is their motto. Separately, birds would no more see such minute fruits than insects would be attracted by the small solitary floret of a Daisy, or the hardly larger flower from an umbel of the wild Carrot. The fact that these small fruits are not only presented in clusters, on a common calyx as on a plate, but that the same shrub or trailing bush bears scores or hundreds of such clusters, is of great advantage to them in the distribution of their seeds ; inasmuch as most of the birds attracted by them feed in flocks, and such an abundant supply has developed the habit among birds, and made it worth their while to stay and dine.

The principle of "mutual help" has perfected, protected, and encouraged the development of plants, as much as it has benefited all the higher members of the animal kingdom, not excepting man himself. We reasonably anticipate that in the future this principle will be infinitely more active in the moral, mental, and physical development of mankind, than its operation amongst ourselves in the past can furnish us with any idea of!

CHAPTER X.

SOCIAL AND POLITICAL ECONOMY OF PLANTS.

IN the animal kingdom the love of offspring has developed the highest traits of character. In man this sentiment, and that of "taking thought for the morrow," have laid the foundation of most social virtues, such as care for others, prudence, and foresight. Nowhere, however, do we find these principles carried out in such a perfect and stereotyped degree as in the vegetable kingdom. We hold life insurance among the middle classes, and registered friendly societies among the working men, to be the outcome of a thoughtful preparation against the future ; and when men toil hard in order to start their children well and fairly in the world, we are disposed to give them generous credit. Both these dispositions, however, are of modern origin—they hardly existed at the beginning of the present century, and are as yet but in their infancy, or there would be less pauperism extant.

Among plants the exercise of the principles both

of prudence and thoughtfulness has been developed to an extraordinary degree; and these principles had attained their stereotyped perfection ages before the appearance of man upon the earth. There is not a single flowering plant which does not leave a *legacy* to its descendants, in the store of nutriment associated with its seeds!

The value of such a vegetable fortune varies to a degree almost human. Some have a rich store, like the embryo of the Cocoa-nut, which feeds on the well-known rich white flesh within the shell, until its radicle penetrates one of the three well-known "monkey eyes" at the end. The germ plants of the Beans, Peas, Vetches, Oaks, Hazels, Walnuts, Brazil-nuts, etc., are also rich in a substantial legacy of food-material, carefully hoarded up by the parent plants in the lobes or *cotyledons*, etc., of the seeds. Even when these lobes are not thick and fleshy as in the Bean and Acorn, they are surrounded by a special provision of albumen and starch foods. Some embryo-plants are well off; others are poor. Those of the Mustard, Cress, Poppy, Nettle, and many other seedling plants, for instance, are provided with so slender a fortune that it is soon exhausted, and the seed-lobes have to develop *chlorophyll* and become green, as we see them when we sow our Mustard-and-cress on damped flannel. If the seed-lobes did not immediately turn to and work like fully developed ordinary leaves, the embryos of such plants would inevitably die, and the species would become extinct. In comparison with the abundant way the young plants of the Oak, etc., are supplied with materials to support them until such time as the radicle can absorb its own mineral food from the soil, and the first leaves of the plumule expand to feed on the carbonic acid in the atmosphere, the speedily-developed green seed-lobes of the Mustard-and-cress are separated as far asunder as the children of wealthy people sent to Eton and Cambridge are from the city arabs who sell fusees in the streets! Still, one cannot but admire the marvellous power of adaptation in these seedling-plants, unpossessed of much food-store, which enables them in lieu of it at once to gain an honest living in another way. To the young plant it comes to the same thing eventually—whether its seed-lobes contain a legacy of nutritious material, or are endowed with speedy and active vegetative energy instead.

There is hardly a single department of the life of most kinds of plants where we do not find the laws of political and social economy in operation, and that to a degree which surprises the student of this science, from the human point of view, and furnishes him with many an apt illustration in the prosecution of his researches.

I have already referred to leaf-buds, and the manner they are protected from frost by those outside and modified leaves called scales or *bracts*. The story of the formation of leaf-buds is not without a moral. People uneducated in botany imagine they are all formed in the spring of the year, but the real

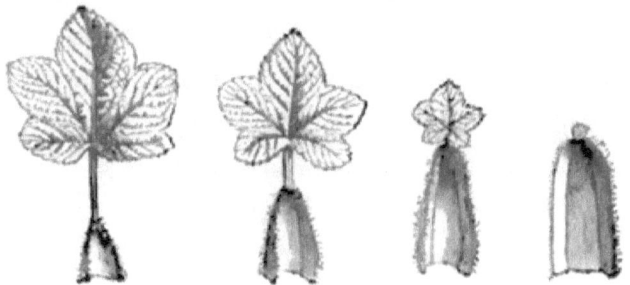

FIG. 70.—Gooseberry leaves gradually passing into scales.

fact is, that before last summer's leaves had fallen their *successors had been already appointed*. The former had not only laboured during the summer to separate the carbon from the oxygen of the carbonic acid gas they had inhaled, they had not only enabled the shrub or tree to add to the store of its woody bulk, but they had further developed the materials out of which their leaf-successors should be fashioned. Leaf-buds are everywhere stores of accumulated or saved up materials, out of which future leaves will be formed. It is always the future rather than the present which is thus kept in view. Some plants, so to speak, bank underground. All they have saved is stored up, not in the form of leaf-buds grouped on the wintry branches, but as underground buds, such as the tubers of the Potato, etc., Artichokes,

Earth-nuts, granules of Saxifrages, Pilewort, etc. These parts of plants are not all roots, as people imagine, but frequently true buds; so that we have underground as well as aboveground buds. From these subterranean parts new plants will arise, as every man knows who has planted potatoes. He is aware that one large potato may be cut up into a number of small pieces, and that each piece will develop, if placed in proper soil, into a new Potato-plant, provided he does not damage the "eyes" in cutting up. These "eyes" are in reality the parts where growth takes place, the rest of the potato being simply so much starch-food, on which the young plant feeds, as certainly as an infant does on its mother's breast, until it gets strong enough to absorb its own nutriment from the soil. Singularly enough, the most beautiful plants in the world resort to this method of underground storage of food-material, all intended either for another season or for another individual. In the "bulbs" of the Hyacinth, Lily, Daffodil, Snowdrop, Tulip, etc., and those of the various Orchids, we have a store of starch, laid by for next year, saved out of last summer's vegetable earnings.

Get the bulb of a Dutch Hyacinth, sold by florists, and place it in the top of one of the coloured glasses made on purpose to receive it, first filling the glass with rain-water, so that the base of the bulb is kept moistened. Let the plant be placed in the window where the sunlight can stimulate its growth.

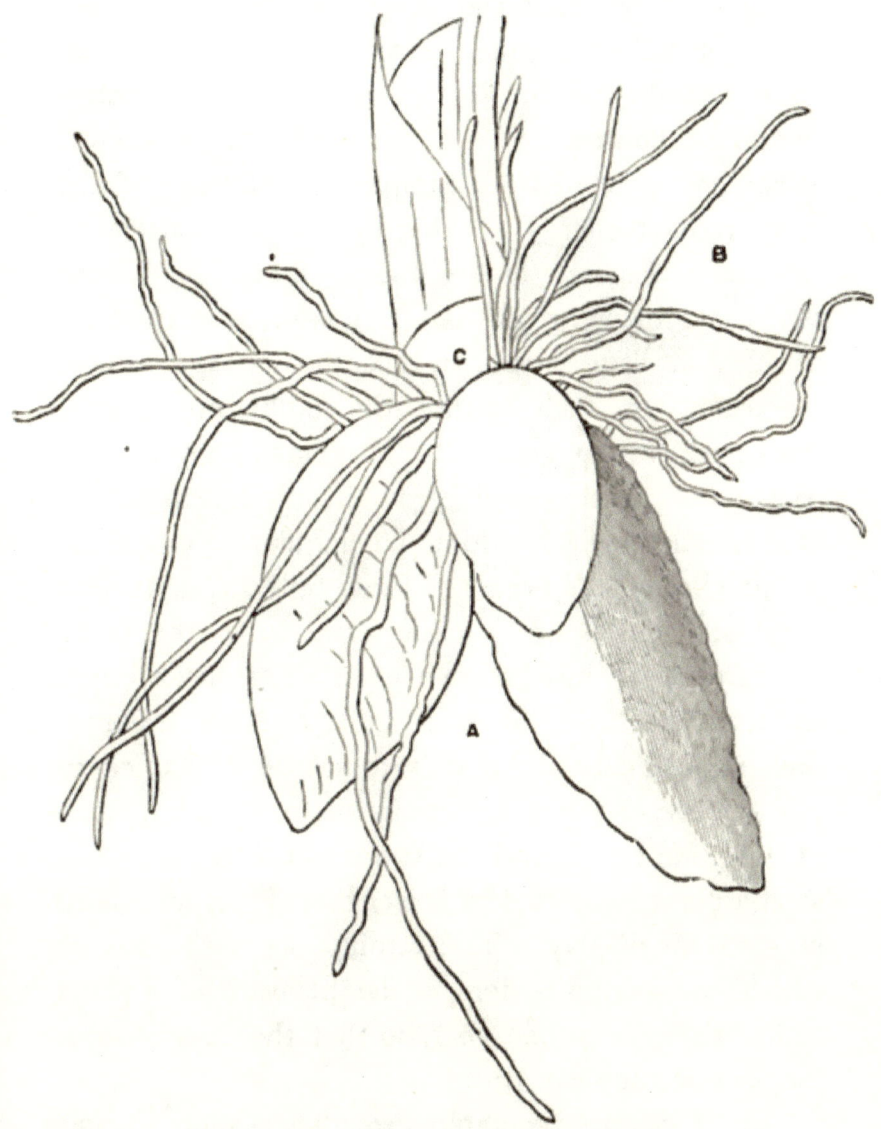

Fig. 71.—Tubers with stored food of *Orchis mascula*.

The bulb asks no more. Water is all it requires,

although we know there is no soil present; and no mineral matter dissolved in the water. The plant develops its bright green leaves, and by and by we have a spike of brilliantly-coloured and sweetly-perfumed flowers. What a transformation scene! What has effected it? Simply the influence of the sun's radiant energy working up the raw material of the store of starch in the almost shapeless bulb, so that the growing plant has been able, with the help of only a little water, to effect a change almost magical. In such plants the bulbs are merely thickened developments of the stems; they are not roots; nor are they underground buds like potatoes. Our horticulturists adopt a singular method of making Lilies and others of these bulbed plants bear very beautiful flowers. They cut down the plant year after year, just before it gives signs of flowering, so that the bulb or base of the stem becomes unusually thickened; that is to say, it is enabled to lay by several years' accumulation of starch instead of only one year's, and the consequence is that when such a plant is allowed to flower there is an unusually magnificent display. This reminds us of the way in which some families forego the pleasure of a short summer's holiday one year, so that they may take a longer one the year after!

Little or no vegetable growth goes on during winter; even though it be a mild one. The plants worked vigorously during the summer, "whilst it

Fig. 72.—Autumn Squill (*Scilla autumnale*), showing large bulb, or store of food-stuff. The reader will notice how much larger is the bulb of the *advanced* Squill with coloured flowers than that of the Garlic with smaller and nearly inconspicuous kinds.

was day," against the dark, cold months, and thus were enabled to store up an excess of food-material, generally in the form of buds or bulbs, after the same fashion that the bears, squirrels, and foxes of Arctic regions get fat just before they retire to sleep the winter through in the hybernating state. Most of our perennial herbaceous plants die down to the ground as the winter comes on; but we shall invariably find that they depend upon next summer's individual resur-

Fig. 73.—Garlic (*Allium vineale*), showing bulb or thickened base of stem where food-stuff is stored up.

rection on the store of starch they laid up in some part or another of their structure underground. They have found out that their succulent stems burst with the winter's frost, and so they protect themselves by storing up recuperative materials below the surface of the soil, where the cold can do them little harm. It is a very severe frost indeed which kills bulbs.

The Primroses, Wild Carrots, Beet, Turnips, and many others, have a root-stock or tap-root, which in some of them actually grows larger every year, something being saved out of the year's annual expenditure for that purpose. Hence the hugeness to which plants of the common Primrose grow in comparison with the small size of the young individuals. Modern agriculture has taken advantage of this

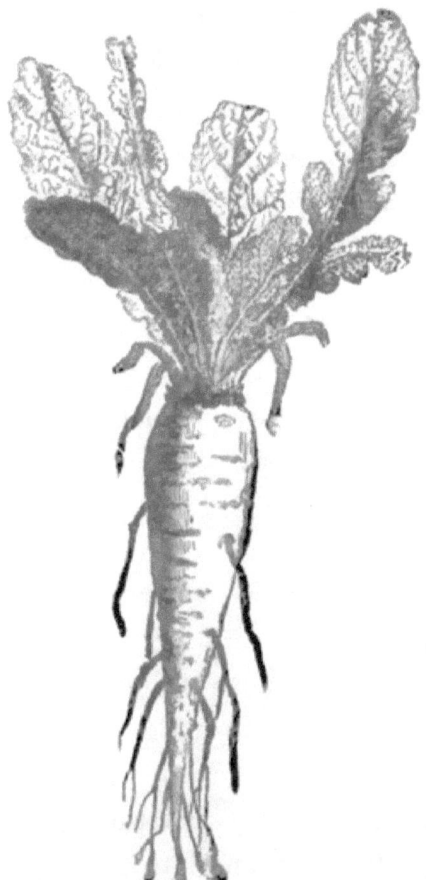

FIG. 74.—Tap-root of Tankard Turnip, showing the root where the food-stuff is stored, and the Rootlets which collect the soluble mineral matter from the soil.

tendency on the part of many plants to store up underground supplies, and has intensified it, and so rendered the results more satisfactory; just as horticulturists have availed themselves of wild fruits, and evolved our larger and more succulent kinds from them. Our Carrots, Turnips, Mangel, Beet, etc., are all artificial enlargements, for purposes of our own, of the parts of plants which store up nutritious food underground because of the rigour of our northern winters.

Not a few plants bury their stems underneath the ground, in which case we know them as *rhizomes*. The Bracken (*Pteris aquilina*) always adopts this plan, although its near relations, the Tree-ferns, lift their fronds high up in the atmosphere by means of thick trunks, which are the exact equivalents of the underground creeping rhizome of the former. But then the Bracken lives where there are cold winters, and the Tree-ferns abound where the winters are very mild. Our Sedges, Iris, and several others, have adopted the same plan as the Bracken, and run their stems beneath the soil, storing and saving up in it all they have been able to accumulate over and above what was necessary to carry on the general expenses. Two very different kinds of organs have thus been specialised to do the same kind of work and to subserve the same economical end, as shown by the rhizome of Solomon's Seal (*Convallaria majalis*), whose popular name is derived from the scars left upon the underground stem by the dead leaves—and its near relatives the Lilies, which store up their vegetable savings in the shape of bulbs.

A distinction must be drawn in this thrifty habit, begotten entirely of climatal conditions. Our Crocuses, Lilies, Daffodils, Hyacinths, Primroses, etc., store up in their bulbs and root-stocks for the benefit of themselves, so that they may not suffer the fate of the feeble little annuals which have not been able to adopt the same plan, and which therefore die when the winter frosts set in. True, whenever excess of food-materials have been elaborated we get bulbils (or buds from the bulbs) formed, so that new individuals are thus generated. But the main thought, so to speak, in their acquirement of this habit of storing up was a *personal* one—that of "providing against a rainy day." The formation of bulbils is an afterthought, and is carried out very unequally.

The Pilewort or Little Celandine (*Ranunculus ficaria*), the Granulated Saxifrage (*Saxifraga granulosa*), the Potato (*Solanum tuberosum*), the Artichoke (*Cynara scolymus*), are less selfish. The individuals die down when winter comes, but the tubers are real underground buds, produced where they are sheltered from mammals and other predatory enemies. The individual plant saves and stores for another generation, as well as for itself. Its patriotism may even induce it to sacrifice its own existence for the benefit of its species. The Strawberry and many other plants adopt the more ingenious plan of merely prolonging their creeping stems. This is a more economical method, and does not involve much saving of plant material—only a little at the swelling of the nodes or joints where the adventitious roots sprout, and where a new plant will shoot.

All bulbous plants, and those thus provided with an abundant supply of easily assimilated foods, have a great advantage over such of their brethren as have not acquired the habit of storing food-stuffs. They are the *first* to avail themselves of the returning light and warmth of the sun, and we find them the characteristic spring and early summer plants. In order to lose neither time nor opportunity, many of them shoot up into flowers at once, like the Snowdrop and Crocus, and so avail themselves of the services of the comparatively few insects which are then abroad, leaving the development of their leaves to the more leisurely opportunities afforded during the later and warmer summer months.

The relation between accumulation and expenditure of organic energy,—the operation of two agencies usually regarded by politicians as completely opposed to each other—is thus seen to be mutually advantageous to the individual plant and also to the species of which it is a member. Reference was made in a previous chapter to such plants as the Aloe and Yucca, which will vegetate for years, meantime storing up abundance of surplus material in their root-stocks, before flowering. Any one who has witnessed the rapidity with which the flowers and spike of the Aloe are developed will see how necessary such a careful and patient husbandry must be to a plant called upon suddenly to expend so much energy upon the act of flowering. One can hardly wonder at the common tradition that the Aloe only "flowers once in a hundred years."

Most species of our native Horsetails (*Equisetaceæ*) have adopted the habit of producing vegetative fronds at one period of the year, and reproductive fronds at another. These are usually so unlike each other that many a young botanist has been led to think them different plants. The stem from which both shoot runs underground, and in its tissues the extra material gained during the vegetative stage is stored until it is required by the reproductive frond. A great number of Ferns have separate fronds: one set purely vegetative and the other as specially reproductive. Our common Hard Fern (*Blechnum boreale*) and Parsley Fern (*Allosorus crispus*) are familiar examples of this principle of division of labour: whilst the Royal Fern (*Osmunda regalis*) has proceeded half-way in the direction of complete differentiation, for its spore-bearing parts occupy the upper parts of fronds whose lower pinnules are true leaves.

Plant economy, however, is confined to no especial set of organs. No botanist doubts that the changes of which these economic modifications are the results, were slowly acquired, so that a long period of time was occupied in producing them. Many plants are living botanical museums, stored with evidences of past organic changes, either in the shape of newly-acquired or of aborted organs. On the whole, however, it will always be found that these alterations have taken place in the interest of the species and with a view to save them from extinction.

Take the well-known order of plants *Geraniaceæ*, in which the genus *Geranium* is chiefly distinguished from *Erodium* by all its species (except the little *Geranium pusillum*,) having ten stamens, whilst the latter has only five perfect ones, although the other five are present, but they are imperfect and useless; in fact they do not bear anthers. Can any one doubt that the Erodiums are merely Geraniums altered by having half their stamens aborted, probably because they were found unnecessary and supernumerary? What is the good of a plant building up useless tissues and organs? Such an act is as wise as that of an already hard-worked man, who deliberately puts heavier burdens on his shoulders.

The importance of accessory parts, such as petals, to all insect-fertilised flowers, has been already shown. But these are modified in a wonderful degree whenever such a change is beneficial—sometimes being enlarged, and occasionally aborted. In the natural history economy of plants there are times when a judicious expenditure proves infinitely more economical, because more advantageous, than niggardly thrift. This principle is in operation among all the Orchids, Lilies, Tulips, Daffodils, Irises, etc. The sub-class to which they belong has been christened *Petaloidea* because of the fact that the calyx and its parts, which in most plants are usually of a dull green, are in them as beautiful both in colour and streak as the petals of the flowers. Such plants have taken the calyx into partnership for floral attractive purposes; and hence we have the most beautiful flowers in the world developed by such a co-operation. It is therefore not without reason the Great Teacher drew attention to the "Lilies of the Field," or declared that "Solomon in all his glory was not arrayed like one of these!"

Many of the plants in another sub-class, that of *Monochlamyds*, bear flowers in which a similar floral modification has taken place, and a *perianth* is the result. Some of these produce very beautiful flowers, as the Daphne, Bistort, etc., although a great number of members of this class are wind-fertilised, and therefore do not require any accessory floral organs. But the latter are often present in a stunted condition, dwarfed occasionally to microscopical smallness. Sometimes the attractive efforts of a flower are aided by one portion only of the calyx being altered for the purpose, as in the

Chalcophyllum, of which Kingsley says: "It is not the flowers themselves which make the glory of the tree. As the flower opens, one calyx-lobe, by a rich vagary of nature, grows into a leaf three inches long, of a splendid scarlet; and the whole end of each branch for two feet or more in length, blazes among the green foliage till you can see it and wonder at it a quarter of a mile away."

Drooping flowers are usually distinguished by having the calyx-lobes either coloured or else shrunken and shrivelled to their extremest tenuity. These modifications, opposite and extreme as they are in their character, nevertheless subserve the same end. It is evident that in drooping flowers the calyx is the most conspicuous part. Therefore, if it is not to interfere with the effect of the coloured corolla, its parts or sepals must be *dwarfed*, as they are in the Common Hairbell (*Campanula rotundifolia*). If they are enlarged they must be *coloured*, and then they act in conjunction with the corolla, assisting its efforts to attract insects, and become beautifully tinted, as in the pendent flowers of the Fuchsia.

Not unfrequently, in flowers where the calyx is the most prominent object, the sepals grow so enlarged that practically the petals of the corolla cannot be seen, or only seen very faintly. This is the condition in which the petals of the Monkshood (*Aconitum napellus*), the Columbine (*Aquilegia vulgaris*), and the Larkspur (*Delphinium ajacis*) now are.

The calyx-parts are highly developed and attractively coloured, and the parts which ought to have performed the work (the petals) are dwarfed and inconspicuous—dethroned, in fact!

The order *Ranunculaceæ* is a very suggestive one in this respect. It abounds in illustrations of the economic and other changes here discussed, and some of its members, such as the *Anemones*, Marsh Marigold (*Caltha palustris*), etc., carry them to the extremest degree. In these flowers the brilliant and glossy red and yellow parts are not petals, as many people naturally suppose, but the sepals of the calyx, which have been promoted to do petal duty; whilst the petals have been dwarfed to mere abortions. The richly supplied *nectaries* will be found at the base of the sepals. A good many Buttercups possess calyx-parts which are said to be *petaloid*, or coloured. Perhaps the reason for so many modifications among the *Ranunculaceæ* is that they are nearly all herbaceous plants.

Both in this order, and more especially in that of the Poppies, we find an understood arrangement that if the sepals cannot help the petals, at any rate they shall not stand in their way. In such cases the sepals fall off as soon as the flower expands, and are termed *fugacious*. In the Tuberous Buttercup (*Ranunculus tuberosus*), and in most if not all of the Poppies, the sepals drop off immediately, leaving the vividly-coloured corollas without a single drawback.

One genus in the order *Ranunculaceæ* manages to do without a corolla, and yet to be a very attractive flower. This is effected by a prominent enlargement of the stamens, whose anthers or pollen-bags are very conspicuous, and as they are very

FIG. 75.—The Water Crowfoot (*Ranunculus aquatilis*)—showing large expanded floating leaves, and thread-like submerged leaves.

numerous and are well clustered together, the clusters of such stamens as seen in the Meadow Rues (*Thalictrum*) form really attractive flowers. In the male flowers of the Willows the stamens are clustered together in a similar but more effective manner, and the male catkins are resplendent in their golden glory in March and April, when the

FIG. 76.—*Ranunculus heterophyllus.*

bees are first out, and eager to find both honey and pollen.

One of the most suggestive facts illustrative of the economic policy of plants is afforded by various species of the genus *Ranunculus*. It is dwelt upon

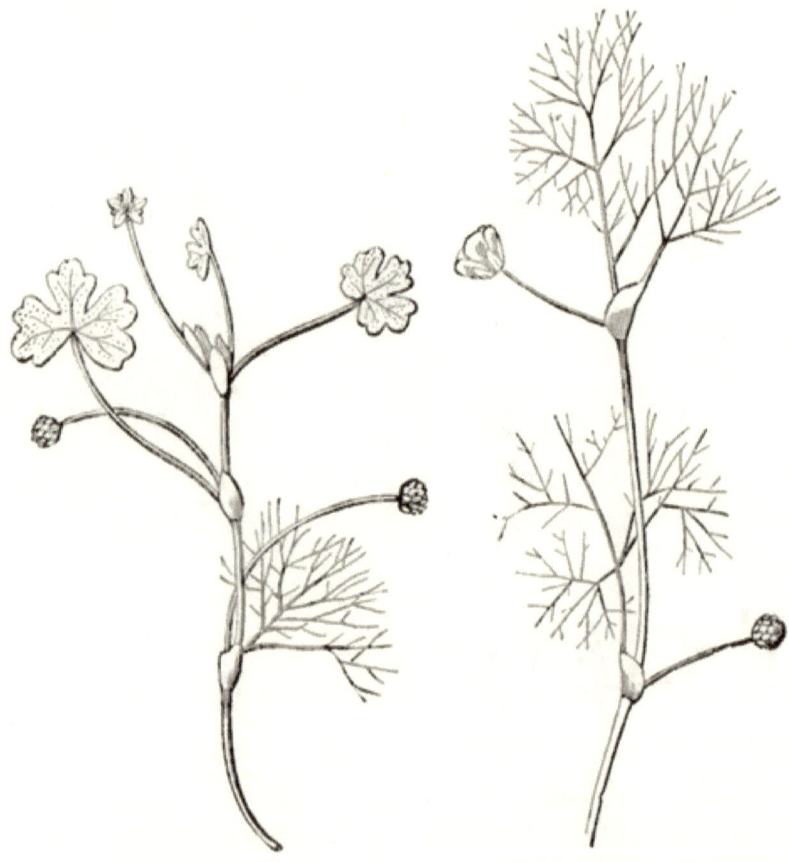

Fig. 77.—*Ranunculus tripartus*, with abbreviated ordinary leaves.

Fig. 78.—*Ranunculus drouettii*, showing entire absence of ordinary leaves.

at length by Mr. Grant Allen; but convenient allusion suggests itself here. Water cannot and does not contain so much carbonic acid as the atmosphere —hence the *filiform* or thread-like shapes of the submerged leaves of water-plants belonging to the most widely separated of orders can be explained by the fact that all are alike subjected to the same physical conditions of poverty, as regards carbonic acid food-supplies. All such submerged leaves are accordingly reduced to a minimum of size, to save vegetal energy on the part of the parent-plants. But several so-called species of more or less aquatic Water-Crowfoots have two or even more sets of leaves, according to physical circumstances. Ordinarily, those of their leaves exposed to the air are expanded in the usual fashion, whilst those submerged are restricted in their development to mere mid-ribs and veins, resembling a cluster of green threads. Our British species of Ranunculuses shows every stage of transition, from the water-feeding to the air-feeding leaves; so that no more interesting genus of plants for evolutionistic purposes occurs in the British flora. (See Figs. 76, 77, 78.)

In one order of plants, the *Euphorbiaceæ*, the usual accessories of a true flower are absent altogether, and mere bracts, or altered leaves, have been pressed into the new service to do the ordinary duty of petals. How well they effect it is seen best of all in the scarlet *Poinsettia*. No Poppy, with the most orthodox of petals, is more colourably prominent. Even our British Spurges, such as *Euphorbia helioscopica*, *E. portlandica*, *E. amygdaloides*, and others, have their flowers resplendent with golden-

Fig. 79.—The Portland Spurge (*Euphorbia portlandica*), showing coloured bracts instead of petals, etc.

green bracts, so that petals are quite unnecessary. The tropical Euphorbias are kept in our greenhouses for the sake of the striking colours which their floral bracts develop. The *Bougainvillea*, belonging to a widely separated order, has evolved a similar method of floral attraction. This plant is now very abundantly grown for the sake of its exquisitely lovely pink flowers; but each of the seeming petals, on examination, prove to be altered bracts or leaves, in which even the venation has not been obscured.

The purpose before me, however, is not to produce an inventory of all the botanical facts bearing upon the political and social economy of plants, but to draw attention to the operations of these principles in the vegetable kingdom.

Nowhere are they better exemplified than among *leaves*. We have already seen that the latter are vegetable units, and their plasticity is evidenced by the readiness with which they can be metamorphosed into carpels, stamens, petals, sepals, bracts, tendrils, or thorns, as occasion may require. This, however, does not interfere with the chief reason of their existence, which is to assist in building up the tissues of the parent-plant or common stock; as well as of preparing both for leaf-successors and plant-continuance.

When leaves are told off to do other work, the plant must make some arrangements by which the duties formerly performed by leaves shall henceforth be undertaken by some other organ or some other part. The work *must* be done—what is to do it? Plants of various orders have found out that the easiest raised substitutes for leaves are *phyllodes*, which are developed by simply *flattening* the branches, leaf-stalks, and flower-stalks, providing them with carbon-feeding mouths (*stomata*), and filling their improvised cells with *chlorophyll* or "leaf-green," so that they can perform all the functions of ordinary leaves, such as decomposing carbonic acid, distilling vapour into dew, and collecting rain. This is the origin of the rigid leaf-like parts of the Butcher's Broom (*Ruscus aculeatus*); the elegant and grass-like leaves of the Grass Pea (*Lathyrus nissolia*), which takes its common name from the fact, although the "grass leaves" are not leaves at all, but *phyllodes* or flattened leaf-stalks; of the Meadow Vetchling (*Lathyrus aphaca*), whose flattened stipules do duty for true leaves, of which the plant has none except on the first petiole; and more especially and universally in the Australian *Acacias*, whose apparent flattened leaves are really *phyllodes*, or modified leaf-stalks, after the fashion just described. Singularly enough, as any one may prove by experiment, the first *real* leaves of the Acacia seedling are compound, exactly like the properly constituted leaves of the Acaciæ of other countries—showing that the Australian branch of the family, for reasons of its own, has modified the original plan of its leaf-morphology. The

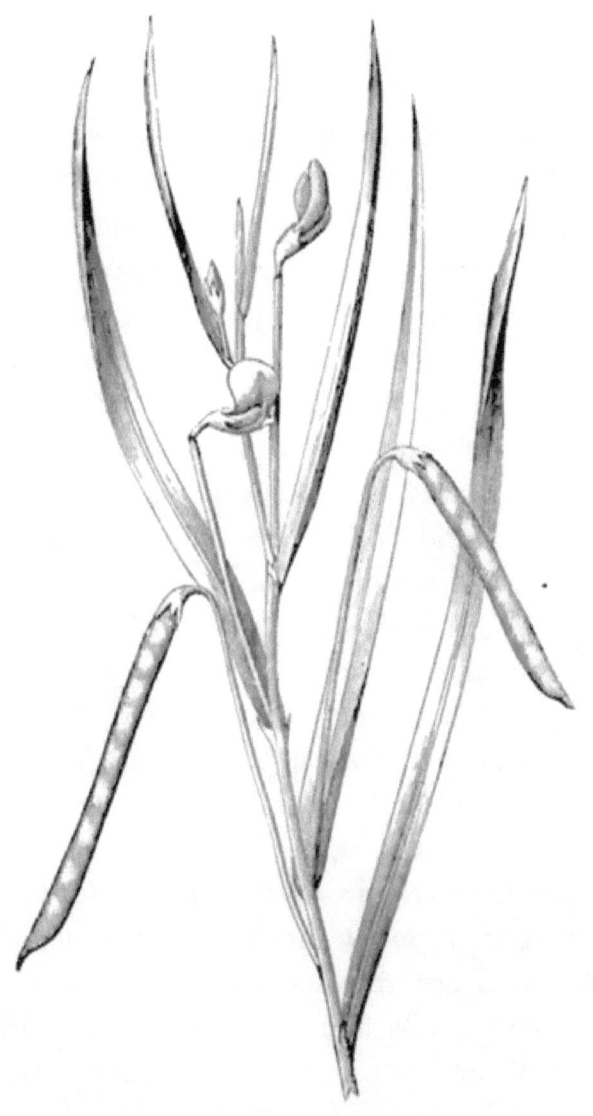

FIG. 80.—Grass Pea (*Lathyrus nissolia*), showing flattened leaf-stalks.

study of the life beginnings both of plants and

animals furnishes the naturalist with a kind of picture, or condensed biography, of the changes through which the species or even genus may have passed. Thus, with the Australian Acacias (as just stated), if we raise a plant from seed, we find that the *first* leaves developed are of the ordinary Acacia kind, with slender petioles and numerous small leaflets; in the next the leaflets are fewer, the main leaf-stalk is dilated and larger in proportion; and so on till the leaflets disappear, and the leaf-stalk has been gradually transformed into the *phyllodium*. The process of transmutation may be seen going on in such plants as *Lathyrus alata*, which bears both *phyllodes* and perfect leaves on the same plants and at the same time.

The Cacti, and those of the *Euphorbiaceæ* which have so singularly mimicked the Cactus-shape and outlines, are always armed with sharp and dangerous thorns to defend the succulent tissues from thirsty and hungry mammals. These sharp, tough spines are in reality leaves, whose original duty has been put off for the more important one of defence. And the functions formerly performed by such transformed leaves are now undertaken by the epidermis of the green trunk and angular stems, which are covered with *stomata*, just as the under surfaces of leaves usually are. In short, the leaf-functions are thus carried out by the whole exposed surfaces of the plant. All those Orchids (chiefly tropical) which are *epiphytal*—that is, have taken to growing upon the bark of trees, so that they can get light, air, and sun at an elevation they otherwise could not possibly hope for, or afford to personally obtain —seldom have many leaves. But, to make up for the lack of organs specially adapted for the work of accumulating energy (doubly important when we remember how lavishly the flowers of Orchids are called upon to expend), the surfaces of the stems, and even of the green *roots*, are usually covered with *stomata*, and thus make up for the deficiency of leaves and small amount of work which would have been done had not some other parts of the plant thus generously come to their aid.

It is possible that many plants are *annuals*—that is, only live a single season, because of the immense number of seeds they are obliged to produce in the interest of the species to which they belong. The latter would be unable to hold its own in the keen specific competition constantly going on unless it was helped out this way, just as the perpetuation of rabbits, rats, and other rodents, depends entirely upon their marvellous fecundity.

CHAPTER XI.

POVERTY AND BANKRUPTCY.

THE subject touched upon in our last chapter opens out a wide field of observation and generalisation. The doctrine of evolution includes "retrogradation" or "degeneration" as well as "progression." This is abundantly verified in the animal kingdom, as Professor Ray Lankester's admirable little volume on the subject demonstrates. Nor is the vegetable kingdom less fruitful in proofs of the same law that, whilst the main mass of living organisms have throughout geological time advanced to higher ground, some have stood still, or merely "marked time," and others have gradually lost ground and dropped out of the ranks. Some of the latter have honestly struggled to keep up the step, and have found themselves unable; others have done so only by diplomacy and cunning; or by the development of qualities even more sinister.

In the vast periods of time which have elapsed since flowering-plants first appeared upon our globe

—comparatively modern though that event is—many of them have undergone similar vicissitudes to those which are nowadays affecting the human inhabitants. Insect-fertilised or *entomophilous* flowers could not of course appear until the flower-haunting class of insects had been introduced, and most probably both were more or less evolved together. Therefore I hold it as reasonable that wind-fertilised or *anemophilous* flowers appeared first in order of time. There is every geological reason to think so, for we have proofs of the most decided of wind-crossed plants, the *Coniferæ*, as far back as the Devonian Period, and therefore many geological ages before the higher *coloured* flowering plants appeared, even allowing they were really in bloom during the Cretaceous epoch, as is not improbable.

But within that vast sweep of time what biological changes may have occurred. Some flowers have been slowly adapted to the visits of insects, and have developed attractive organs accordingly. Ages have passed away; other flowers of a higher organisation have competed for the mastery; other insects have been evolved. Some of the older flowering-plants have had to give way slowly, inch by inch, as it were, and retire from their exalted *entomophilous* habit to the more ancient *anemophilous* mode of life from which they originally sprang! Theirs has been a brave but a losing fight. They have now retired from the higher ranks—they are floral bankrupts. Traces of the higher floral rank enjoyed by their ancestors may still be found in their flowers, as in the *lodicules* or inner scales of the wind-fertilised flowers of Wheat, many species of Grasses, in the Wood-rushes (*Luzula*), and Juncaceæ, as *Juncus bufonius*, for instance, which are all that remain of the original petals; just as a poor but proud and ancient family cling to their "crest" and coat-of-arms long after their ancestral estates and property have disappeared. Changes

FIG. 81.—The Toad Rush (*Juncus bufonius*).

from the entomophilous to the anemophilous habit of life, and contrariwise, have constantly been going on in flowers, pendulum-like, from one extreme to the other, according as the change has been necessary or beneficial, or preserved species from total extinction.

Every botanist is aware that many plants not only bear fewer sepals and petals than they ought, according to the typical structure of their flowers, but fewer carpels as well. *Five* is the normal number of stamens and pistils in all dicotyledonous flowers, and *three* in all monocotyledonous kinds. These two typical numbers are fairly adhered to in both divisions, but there are numerous instances indicating that the decreased number of floral organs is due to suppression—in short, to floral poverty. Thus the Saltworts (*Salicornia*) and Marestails (*Hippuris*) have only one stamen and one pistil; the Starworts (*Callitriche*) one stamen, but two pistils; the pretty little Veronicas or Speedwells, on the contrary, have two stamens and only one pistil; as is also the case with the Privet, Salvia, etc. The Burnets, Plantains, Teasels, Scabious, etc., have only four stamens. In these flowers the petals are usually only *four* in number. Practically, all the departures from the number five are due to suppression of the parts lacking to make up that number; and we may reasonably assign as a cause for such reduction, the necessity experienced by the plant for reducing its floral expenditure. Sometimes we catch a plant in the act of suppression. Thus, if we examine the flower of the Oak we find it has *two* ovaries, each containing ovules. But only *one* of these comes to anything—that is, to an acorn. The other, instead of increasing in size, decreases and dies. The carpels of many flowers are suppressed in a similar way during maturation. The plant finds it could not afford to part with the albuminoid materials necessary to furnish such a supply, and so it reduces their numbers. The Oak has reduced its normal number of five carpels to two, and is now engaged in still further limiting them to one.

In other cases a different plan is adopted for limiting the number of seeds, especially when the latter are large, and therefore likely to drain the parent-plant of a considerable amount of food-stuffs, difficult to be replaced, as everybody who has observed is aware how an Apple-tree is weakened after bearing an unusual quantity of fruit. Few trees have more brilliant flowers than the Horse-chestnut, but only some of each spike appear to be successfully fertilised. The fruits which subsequently appear are still numerous, but afterwards commences the "slaughter of the innocents." The law of natural selection goes on in each cluster of young chestnuts. The tree cannot possibly afford to supply them all with food-stuffs. They wither and droop, and cover the ground with blighted immature fruits. Those hidden from sunshine go first, for they are unfavourably situated. Weak-stemmed and inefficiently-crossed individuals follow, and so on; until only one or two prickly husks remain to fully ripen, where perhaps a dozen commenced, all seemingly with the same chances of success!

Grant Allen (*Flowers and their Pedigrees*) holds that Wheat and Grasses are florally degraded Lilies, whose three carpels are reduced to one, and that with only a solitary seed in each instance. The *lodicules* (or small inner scales seen in the flowers of the Wheat and most Grasses) are all that remain of what were formerly petals or perianth-segments. In *Scirpus* these parts are represented by six small dry bristles. The Carices have lost all traces of any former perianth. Some Grasses have reduced their normal number of stamens from three to two, as the Vernal-grass (*Anthoxanthum odoratum*), for instance; and numbers of them have lowered the number of their stigmas to the two plumose parts everybody is acquainted with.

Aborted male and female organs (stamens and pistils) occur in the clusters on the spadix of the well-known "Lords and Ladies" (*Arum maculatum*) and in other British plants. In every one of these instances we are warranted in regarding such an occurrence as evidence of degeneration.

The successful habit the "Cleavers" (*Galium aparine*) has adopted, of climbing by means of multitudes of grapnel-hooks, has already been alluded to. The minute white flowers of that plant, however, should not fail to be remarked. They spring in feeble clusters from the verticillate or star-like arrangement of leaves, and are anything but attractive. And yet a simple dissection shows they belong to that highly-organised group of flowers in which all the petals have become united in a single piece (*gamopetalous*). The family to which the "Cleavers" belongs possesses some very brilliant members, and it does not require much research to see that the Cleavers has degenerated ; that once in its individual history it possessed much larger and perhaps more brilliant flowers, fertilised by the higher insects. Now it is so poverty-stricken it can only produce these small ones, which are fertilised by insects diminutive in proportion. The near allies of the Cleavers, the Bedstraws (*Galium mollugo, Galium verum*, etc.), have also suffered floral reverses, but they to some extent make up for them by adopting the social or co-operative principle, and cluster their small flowers in spikes and tufts, so that the hedgerows are gay with their white and yellow blossoms. The pretty Field-madder (*Sherardia arvensis*) still retains the colour of its high rank—a lilac tint ; and both this and its shape proclaim that our feeble little herb, hiding away in our cornfields, is another illustration of both floral and vegetal degeneration. The fact that all the above plants have only four, instead of five petals united to form their corollas, is another bit of evidence pointing to the same conclusion. The tropical members of this order are many of them possessed of very beautiful flowers, such as the *Gardenias*, *Coffea*, etc., and assume arboreal proportions and magnitude, as witness the *Cinchona* trees. All our British species are " poor relations" of this otherwise highly-developed group, and botanical evidence proves that they "have come down in the world."

The variability of flowers, extending through every degree, from minute kinds like those of our Field Speedwell to the large Field Poppy, represents the relative success which has accompanied the efforts to attract insects. What a large range of variation occurs in this respect in every order, and even in some genera, as in that of *Geranium*. The same natural group often contains both wind-fertilised and insect-fertilised plants. These facts plainly indicate the mutabilities to which nearly-allied species have been subjected.

Another feature should also be noticed—the inequality in the duration of the *individual* lives of plants. It is almost as diverse as the sizes and colours of flowers. Some are *annuals*, others *biennials*, and the most favoured are *perennials*, or else woody shrubs and trees whose duration is very great. Human life and fortune are not more capricious than the fate which frequently governs the various members of the same genus of plants. Take the Buttercups (*Ranunculaceæ*). Why should *Ranunculus repens*, *R. acris*, and *R. bulbosus* be "perennial," whilst the nearly-allied species *Ranunculus parviflorus* and *R. arvensis* are "annuals"? It is not because the latter exhaust themselves in the act of inflorescence, for their flowers are smaller than those of their more favoured brethren. One can understand why the Poppies should be annuals, for they expend all their capital in advertising—that is, in the development of large petals—as well as in the enormous amount of pollen they form (more than three and a half million grains in each individual), and in the immense number of small seeds. Such a drain on the vitality of a plant in so short a period of time as a single summer—such an abstraction of phosphorous and nitrogen for pollen-grains, and of albuminous material for seeds—would tend to render even a large plant bankrupt. But the Welsh Poppy (*Meconopsis cambrica*) is "perennial," although its brilliant yellow petals are nearly as large as the red ones of our field species.

It is worth our while to observe the unequal duration of the lives of plants collected into the same genus, inasmuch as it must be due to certain habits in the plant's present economy, or is the effect of accumulated habit in the lives of ancestors. Perhaps in a general way it can be best accounted for by some plants being more thrifty than others—that is, they have learned to accumulate more than they expend; and their excess is laid up underground in a thickened stem, like *Ranunculus bulbosus*, or in root-stalks, rhizomes, and the like. There is even a variation in the brief lives of "annuals." Some extend through the entire summer; others shoot up, flower, and seed, all in the space of a few weeks, as if the most important thing in their lives was to die and get out of the way as soon as possible!

The *size* to which the entire plant grows is subject

to the same variability; and this is, perhaps, more notable in the woody kinds than in herbaceous, although it is abundantly exemplified in the latter. Of course we cannot tell to what dimensions the original ancestor of any plants grew; and doubtless we should find that the progeny of all have experienced varying successes and reverses like those the human descendants of Charlemagne and Alfred have gone through, if we could formulate them. Some may have attained a huger bulk than their forerunners, and probably this is usually the case; but many others have retrograded. Still, the number of the latter cannot exceed that of the former, any more than paupers can form the bulk of the population. These botanical waifs and strays may even throw some light on the evolution of their kind, and picture to us the physical experience they have undergone; as witness the order of the Willows (*Salicaceæ*), even within the bounds of the genus *Salix*, all of whose species must have had a common origin. These members vary from the White Willow (*Salix alba*—Constable's *Queen of the Meadows*), seen so abundantly in our river-valleys, to the Least Willow (*Salix herbacea*), found only on the bleak tops of the Scottish mountains, where it never exceeds the length of six inches! What a story does this difference in size tell of opposite conditions of life—the dwarf species fighting it out on the bleak and exposed mountain-top, amid frosts and snows and poverty of mineral food—whilst the other luxuriates in the warmer temperature of sheltered river-valleys, where its roots ramble through the richest of alluvial soils to never-failing supplies of water!

The variation in the personal fortunes of members of the same clan is even better illustrated by the British members of our order *Rosaceæ*. If some of them do not grow here to the size of trees, they get as near as they can, like our Wild Apple, Pear, Cherry, Plum, and Mountain Ash, closely followed by the still varying species of the true Roses on a smaller scale, and the unstable host of our creeping Brambles. Most of these are splendidly adapted, in their flowers, to insect-fertilisation, whilst the high degree in which they have evolved fruits— that is, have grown layers of sweet and pulpy matter, often highly coloured and attractive, around their seeds—has enlisted the services of birds to disseminate them. Compare all these evidences of high vegetable specialisation, and then turn to those diminutive members of the same order, the Ladies' Mantles (*Alchemillæ*), one species of which is a puny annual, picking up a few months' scanty summer life as a weed in our waste ground, whilst another has assumed an *Alpine* habitat—has withdrawn, as many human fugitives have done in times of danger and peril, to the mountain slopes and fastnesses. Even their prettily-cut leaves cannot conceal their floral and frugiferous poverty; neither their flowers nor fruits have any beauty that any insect or animal should desire them.

Among the Cryptogamia a similar fickleness is displayed. Between the giant Tree-ferns, growing to the height of 40 and 50 feet, and the Filmy Ferns (*Hymenophyllum*), some of which rarely grow to more than two or three inches, we have a range of size and mass hardly exceeded in any other order of plants. And yet both these kinds of Ferns pass through almost identically the same embryonic stages. Both spring from the fertilised archegoniums produced on prothalli, so nearly alike in size and shape that it seems wonderful how one can develop into a forest tree, and the other remain to compete for possession of moist ground with diminutive Mosses!

There are few orders of plants which do not possess "poor relations"; some are even worse off, for they have to own kinship, not only with "doubtful characters," but with actual robbers and murderers of their own kind, as the Convolvulus has with the Dodder. Whilst the main mass of the members of an order have progressed, or at any rate held their own, a few have slowly fallen behind—have become smaller of size, feebler of stem, possess fewer flowers (and those gradually dwarfed), until eventually extinction terminates a struggle which, perhaps, has extended through an entire geological period.

Many living plants even now are contending in

this downhill fight. They cannot hold their own—they find circumstances too much for them. Accordingly we find they are cutting down their expenditure. Some have done this until there is little or nothing left to economise further. Frail annuals, whose entire mass, perhaps, does not weigh half an ounce, find it impossible to propagate their kind without resorting to various devices, and assuming habits as widely different from those of their better-off kinsmen, as frequenting the pawnshop by honest but needy gentlewomen contrasts with the afternoon airing in the Park of their wealthier cousins.

"The poor ye have always with you" might have been addressed to the vegetable kingdom instead of the human family; and it is of these I like especially to speak. We have seen that many species of plants, not confined to any particular order, have found it difficult to keep pace with life, and that they have all alike adopted a simple method by which extra expenditure could be prevented. In many species this has proceeded so far as to evolve a peculiar set of flowers, called *Cleistogamic*, which may be regarded as ordinary flowers arrested in growth, so that they never open. In short, they never attain to a higher rank than closed flower-buds, instead of developing and expanding into true flowers. As if to show us that this singular group of flowers has come into existence through poverty, we find them in every degree and stage of abortion, from a mere pin's head in size to others which have grown to the very point of bursting open. Further, the habit appears to be dropped and acquired according to circumstances; as is shown by the Yellow Pansy, which has no cleistogamic flowers in its native hilly habits, but develops them when brought into the plains. The cleistogamic habit is not confined to any particular order of plants, although herbaceous kinds are those which chiefly adopt it. We can only explain its origin as an act of floral bankruptcy.

It has already been pointed out that some plants bring forth flowers which are habitually self-fertilised, in which case the flowers are always small and inconspicuous. There is little doubt this resort to self-fertilisation is an act of poverty—an inability on the part of the plant to expend much energy on inflorescence. Compare the Common Groundsel (*Senecio vulgaris*), a miserable and unattractive little annual—fortunate in still possessing the mechanical contrivance (*pappus*) peculiar to its order, or it would soon be extinct—with the Ox-eye Daisy (*Chrysanthemum leucanthemum*) and Corn Marigold (*Chrysanthemum segetum*), or the larger Sun-flower. The Groundsel has lost those external ray-florets which give to the latter plants their greatest attraction. A few of the near relations of the Groundsel, such as *Senecio viscosus* and *Senecio squalidus*, are little better off in this respect, and appear to be on the downward path.

From a self-fertilised habit, acquired under these circumstances of floral squalor, it is but a step forward to the development of cleistogamic flowers. In them there is no possible waste of pollen, and therefore very few grains are sufficient, ranging from about 400 in those of the Wood Sorrel (*Oxalis acetosella*), whereas the gorgeous flowers of the wild Pæony (*Pæonia officinalis*) it has been calculated, produce no fewer than 3,650,000 pollen-grains. Our Wheat-plants, when in flower, yield about fifty pounds of pollen to the acre. Hence the necessity for richly manuring cornlands, to supply the phosphorus and nitrogen thus carried off. In Canadian forests the ground is often thickly strewn with pollen from the Pines, which drifts along the ground, fills up hollows, and lies on the surface of still waters and lakes like a crust. The ground beneath the Spanish Chestnuts in our gardens is often completely covered during their flowering season.

The Palms are quite as prodigal of their pollen as the Pines. Gosse (*The Naturalist in Jamaica*) describing the well-known Mountain Cabbage Palm (*Areca oleracea*), common in certain places in Jamaica, and one of the noblest of all this stately tribe, says: "The immense spike of blossom that projects in the early autumn from the base of the crown, arching gracefully downwards, is a fine object. I have seen, at such times, the earth beneath the tree, for a space of many square yards, quite white with the scattered pollen, as if a light snow shower had fallen."

Such an enormous sacrifice of detached substances, rich in the elements plants find it hardest to obtain, can only be afforded by comparatively few. By far the greater number have to be content with less expenditure, and some are even nowadays slowly falling into arrears, until the cleistogamic habit will have to be adopted to save them from utter extinction. By hoarding up every grain of pollen, self-fertilisation is ensured, and seeds are thus produced in tolerable abundance. As Darwin says (*Forms of Flowers*): "Cleistogamic flowers afford an abundant supply of seeds with little expenditure, and we can hardly doubt that they have had their structure modified and degraded for this special purpose."

The degree to which this degradation of the usual floral organs is carried out in cleistogamic flowers varies considerably. Some Violets have suppressed three out of the five of their stamens, so that only two now produce pollen within the "closed walls" of the never-opening flowers. We also find the petals of cleistogamic flowers in every stage of abortion. Some are almost perfect, as in the cleistogamic flowers of the Grass Pea (*Lathyrus nissolia*); in others, as the Violets, we have a mere trace, whilst in the closed flowers of the Wood Sorrel and White Dead Nettle they are entirely obliterated.

But in this bitter fight with poverty there is a

touching episode savouring of humanity. As much of the old *show* is kept up as the plant can possibly afford, and there are few species which do not bear ordinary flowers as if nothing were the matter; whilst the dwarfed and aborted cleistogamic flowers are hidden out of sight at the bases of the clustering leaves, as though the plant were anxious they should not be seen. The best face possible is put on the case, and often not without good results, for the occasional crossing the conspicuous flowers of these plants get enables the seeds to gain back some of their old vigour, or to stay off the evil days of extinction in which pure cleistogamism might end. The conspicuous flowers are not borne every year by some plants—they cannot afford such a luxury. And one or two known kinds bear flowers which are of no good whatever, for they are never found fertile; so in their case we must regard the habit as a survival, or as an indisposition to give up the old floral life and rank.

The story of the common Red Clover, imported into New Zealand, is an interesting one, as showing how flowers change their habits under certain circumstances. My readers will frequently see it stated in all books dealing with the fertilisation of flowers, that the Red Clover does not bear seed in New Zealand, owing to the absence of humble bees. Mr. J. B. Armstrong, of the Christchurch Botanic Garden, has recently contradicted this statement, and has shown there are four varieties of Red Clover in New Zealand, all of which produce seeds of good germinating power. One variety is partly self-fertile and partly self-sterile. The produce of those plants which have been grown in the colony for several generations tends almost invariably to become self-fertilising. Mr. Armstrong thinks there is every reason to believe that the Red Clover is also becoming modified in its structure, so as to admit the visits of insects not known to visit it in England; and that such modification tends to render the plant self-fertilising, but at the same time enables it to be improved in constitutional vigour by occasional inter-crossing.

The seeds produced by so much hard thrift are cared for and protected, just as devoted mothers would protect the scanty meals of their offspring. Darwin expresses the ingenuity with which they are concealed from the greedy eyes of birds or other seed devourers: "It is one of the many remarkable peculiarities of the plants which bear cleistogamic flowers, that an incomparably larger proportion of them than of ordinary plants *bury their young ovaries in the ground*—an action which, it may be presumed, serves to protect them from being devoured by birds or other enemies." Of course, as he also remarks, they have to sacrifice all those advantages of wind or animal agencies in their dissemination that we have already noticed; but it is nothing to the poor to be obliged to sacrifice — their whole lives are frequently one act of surrender, or giving up the gratification of desires.

"Adversity is the mother of invention"—especially when the invention is to be applied for the benefit of offspring. Hence we are not surprised, knowing as we do how the cleistogamic habit has been acquired and the circumstances which have led to its adoption, to find Darwin speaking as follows: "Cleistogamic flowers possess great facilities for burying their young ovaries or capsules, owing to their small size, pointed shape, closed condition, and the absence of a corolla; and we can thus understand how it is that so many of them have acquired this curious habit."

"Poverty makes acquaintance with strange bedfellows," and it is suggestive to find there is such a thing as floral bankruptcy, resorted to by insect-fertilised and wind-fertilised plants alike; and by members of orders botanically as far apart as possible, but which meet, as peer, merchant, and shopkeeper among ourselves frequently do, in a common Bankruptcy Court.

CHAPTER XII.

ROBBERY AND MURDER.

IT is a difficult matter in analysing the various sensations produced in our minds by studying the habits of plants and their organs—all of which illustrate some of those great principles of Human Conduct around which the crises of history have revolved—to sufficiently separate one class from another by well-defined lines. Degrading poverty so often leads with ourselves to crime that we cannot wonder if the same inflexible condition has caused reduced and pauperised plants to resort to a parasitic life—that is, to live by preying upon others, to the detriment and even death of the latter.

The most remarkable fact which strikes the botanist as he approaches the study of plants from this side is the varying degrees in which parasitism prevails, and how the habit has been indulged in by plants whose relationships are as wide asunder as possible. One chief cause for their assuming this mode of life suggests itself—inability to compete on equal terms with other plants. Hundreds of species are running a scratch-race for life, and are only just able to keep up with it. Once a species falls behind its chance is over, if it contains no store or stock of energy to enable it to gain back the lost ground, and to rush abreast again with a "spurt." *Degeneration* then of necessity follows, and with it all sorts of vegetable vices and dodges to gain a bare living, or for hanging on to life.

The term "parasitism" is often very loosely employed. Many people speak of climbing plants as if they were parasites; but such a condition of vegetable existence is one of *dependency* rather than of *parasitism*. Nevertheless, the act of climbing, or growing in company with some other and stronger plant, may range from "commensalism" to absolute strangulation. One habit does no harm whatever, and may even be a serviceable companionship (for we know nothing as yet of vegetable commensalism, and very little of animal, in spite of Van Beneden's delightful book on the subject); the latter is an act of robbery and murder. Kingsley (*At Last*) gives an instance from his Trinidad experience of companionship in growth which seems almost to come under the head of "commensalism" or vegetable "messmateship." "A *Poix doux* (Inga), some said it was; others that it was a Figuier (*Ficus*). I incline to the former belief, as the leaves seemed to me pinnated; but the doubt was pardonable enough.

There was not a leaf on the tree which was not 100 feet over our heads. For size of spurs and wealth of parasites the tree was almost as remarkable as the Ceibæ. But the curiosity of the tree was a Carat Palm, which had started between its very roots; had run its straight and slender stem up parallel with the bole of its companion, and had then pierced through the head of the tree, and all its wilderness of lianes, till it spread its huge flat crown of fans among the highest branches more than 100 feet aloft. The contrast between the two forms of vegetation, each so grand, but as utterly different in every line as they are in botanical affinities, and yet both living together in such close embrace, was very noteworthy—a good example of the rule that while competition is most severe between forms most closely allied, forms extremely wide apart may not compete at all, because each needs something which the other does not."

In our English climate our intensest conditions of vegetable life are not to be compared with its luxuriant growth in the tropics, especially where humidity is favourable. Burmeister has left it on record that the contemplation of a Brazilian forest produced on him a painful impression, on account of the vegetation displaying a spirit of restless selfishness, eager emulation, and craftiness. H. W. Bates (*The Naturalist on the River Amazons*) has the following remarks: "In these tropical forests each plant and tree seems to be striving to outvie its fellow, struggling upwards towards light and air—branch, leaf, and stem—regardless of its neighbours. Parasitic plants are seen fastening with firm grip on others, making use of them with reckless indifference as instruments for their own advancement. Live and let live is clearly not the maxim taught to us in these wildernesses. There is one kind of parasitic tree, very common near Pará, which exhibits this feature in a very prominent manner. It is called the Sipó Matador, or the 'Murderer Liana.' It belongs to the Fig order. The base of its stem would be unable to bear the weight of the upper growth; it is obliged, therefore, to support itself on a tree of another species. In this it is not essentially different from other climbing plants and trees; but the way the Matador sets about it is peculiar, and produces certainly a disagreeable impression. It springs up close to the tree on which it intends to fix itself, and the wood of its stem grows by spreading itself like a plastic mould over one side of the trunk of its supporter. It then puts forth from each side an arm-like branch, which grows rapidly, and looks as though a stream of sap were flowing and hardening as it went. This adheres closely to the trunk of the victim, and the two arms meet on the opposite side and blend together. These arms are put forth at somewhat regular intervals in mounting upwards, and the victim, when the strangler is full-grown, becomes tightly clasped by a number of inflexible rings. These rings gradually grow larger as the murderer flourishes, rearing its crown of foliage to the sky mingled with that of its neighbour, and in course of time they kill it by stopping the flow of its sap. The strange spectacle then remains of the selfish parasite clasping in its arms the lifeless and decaying body of its victim, which had been a help to its own growth!

"The Murderer Sipó merely exhibits, in a more conspicuous manner than usual, the struggle which necessarily exists amongst vegetable forms in these crowded forests, where individual is competing with individual, and species with species, all striving to reach light and air in order to unfold their leaves and perfect their organs of fructification. All species entail in their successful struggles the injury or destruction of many of their neighbours or supporters; but the process is not in others so speaking to the eye as in the case of the Matador. The efforts to spread their roots are as strenuous in some plants and trees as the struggle to mount upwards is in others. From these apparent strivings result the buttressed stems, the dangling air-roots, and other similar phenomena. The competition among organised beings exists everywhere, in every zone, in both the animal and vegetable kingdoms. It is doubtless most severe, on the whole, in tropical countries; but its display in vegetable forms in the forest is no exceptional phenomenon. It is only more conspicuously exhibited, owing perhaps to its affecting principally the vegetative organs —root, stem, and leaf—whose growth is also stimulated by the intense light, the warmth, and the humidity."

Numerous accounts of the conditions of growth in tropical forests, all penned by able and observant naturalists, exist in our literature; and it is curious to see how little they vary, notwithstanding the strong individualities of the writers. Dr. A. R. Wallace (*Travels on the Amazon and Rio Negro*) says: "At about two miles from the city we entered the virgin forest, which the increased height of the trees and the deeper shade had some time told us we were approaching. Its striking characteristics were the great number and variety of the forest-trees, their trunks rising frequently for 60 or 80 feet without a branch, and perfectly straight; the huge creepers which climb about them, sometimes stretched obliquely from their summits like the stays of a mast, sometimes winding around their trunks like immense serpents waiting for their prey. Here, two or three together, twisting spirally round each other, as if to bind securely these monarchs of the forest; there, they form tangled festoons, and, covered themselves with small creepers and parasitic plants, hide the parent stem from sight."

Mr. Bates (*The Naturalist on the River Amazons*)

describes almost his first impressions of the forest scenery in the same region as follows: "The leafy crowns of the trees, scarcely two of which could be seen together of the same kind, were now far away above us—in another world, as it were. We could only see at times, where there was a break above, the tracery of the foliage against the clear blue sky. Sometimes the leaves were palmate, or of the shape of large outstretched hands; at others, finely cut or feathery like the leaves of Mimosæ. Below, the tree-trunks were everywhere linked together by Sipós; the woody, flexible stems of climbing and creeping trees, whose foliage is far away above, mingled with that of the taller independent trees. Some were twisted in strands like cables; others had thick stems contorted in every variety of shape, entwining snake-like round the tree-trunks or forming gigantic loops and coils among the larger branches; others, again, were of zigzag shape, or indented like the steps of a staircase, sweeping from the ground to a giddy height.

"It interested me much afterwards to find that these climbing trees do not form any particular family or genus. There is no order of plants whose especial habit is to climb; but species of many—and the most diverse families, the bulk of whose members are not climbers—seem to have been driven by circumstances to adopt this habit."

Lastly, Mr. P. H. Gosse, thus gives his experience

of the forest in Jamaica: "A steep rocky hill rises abruptly, covered with pristine woods. The boughs of an immense Fig-tree, which had been prostrated in a storm a few weeks before, enabled me to climb the ascent; but I was astonished at the difficulty of penetrating the forest. The numbers of tough withes, many of them fearfully spinous, that entwine about the trees and about each other; the long prickly *Cacti*, too, that trail here and there; the lianes, that resemble ropes, or lines, or strings, according to their thickness, hanging down in loops, or loosely waving to and fro—are wonderful; these last frequently extend from a lofty bough nearly to the ground, without a branch or leaf till near the extremity, where the cord commonly divides into three or four more slender ones. Some of the larger ones are woody, and are often seen tightly twisted together, like the strands of a cable."

I shall conclude these varying but coincident references to the vegetable strife and competition raging in tropical forests, both in the Old and New World, by the following suggestive paragraph from Kingsley's *At Last:* "As we proceeded we entered a forest still unburnt, and a tangle of beauty, such as we saw at Chaguanas. . . . Overhead sprawled and dangled the Common Vine-Bamboo (*Panicum divaricatum*), ugly and unsatisfactory in form, because it has not yet, seemingly, made up its mind whether it will become an arborescent or a climbing grass;

and, meantime, tries to stand upright on stems quite unable to support it, and tumbles helplessly into the neighbouring copsewood, taking every one's arm without asking leave. A few ages hence its ablest descendants will probably have made their choice, if they have constitution enough to survive in the battle of life—which, from the commonness of the plant, they seem likely to have. And what their choice will be there is little doubt. There are trees here of a truly noble nature, whose ancestors have conquered ages since; it may be by selfish and questionable means. But their descendants, secure in their own power, can afford to be generous, and allow a whole world of lesser plants to nestle in their branches, another world to fatten round their feet. There are humble and modest plants, too, here—and those some of the loveliest—which have long since cast away all ambition, and are content to crouch or perch anywhere, if only they may be allowed a chance ray of light and a chance drop of water wherewith to perfect their flowers and seed. But, throughout the great republic of the forest, the motto of the majority is—as it is, and always has been, with human beings—'Every one for himself, and the devil take the hindmost!' Selfish competition, over-reaching tyranny, the temper which fawns and clings as long as it is down, and when it has risen, kicks over the stool by which it has climbed—these, and the other 'works of the flesh,' are the works of the average plant, as far as it can practise them. So by the time the Bamboo-Vine makes up its mind, it will have discovered, by the experience of many generations, the value of the proverb—'Never do for yourself what you can get another to do for you,' and will have developed into a true high climber, selfish and insolent, choking and strangling, like yonder beautiful green pest, of which beware—namely, a tangle of Razor-grass (*Scleria flagellum*). The brother, in old times, of that broad-leaved Sedge which carries the shot-seeds, it has long since found it more profitable to lean on others than to stand on its own legs, and has developed itself accordingly. It has climbed up the shrubs some fifteen feet, and is now tumbling down again in masses of purest deep green, which are always softly rounded, because each slender leaf is sabre-shaped, and always curves inwards and downwards into the mass, presenting to the passer thousands of minute saw-edges, hard enough and sharp enough to cut clothes, skin, and flesh to ribbons, if it is brushed in the direction of the leaves. For shape and colour few plants would look more lovely in a hot-house; but it would soon need to be confined in a den by itself, like a jaguar or an alligator!"

Even more to the point raised in this chapter is Kingsley's reference to a confirmed strangler: "He will look up, with something like a malediction, at the Matapolo, which, every fifty yards, have seized on mighty trees, and are enjoying, I presume, every different stage of the strangling art, from the baby Matapolo, who, like the one which you saw in the Botanic Garden, has let down his first air-root along his victim's stem, to the old sinner whose dark crown of leaves is supported eighty feet in air, on innumerable branching columns of every size, cross-clasped to each other by transverse bars. The giant-tree on which his seed first fell has rotted away utterly, and he stands in its place, prospering in his wickedness, like certain folk whom David knew too well!"

Some of these parasites assume the habit only when convenient, and they appear to be able to throw it off as may be required. The genus *Clusia*, abundant in the forests of tropical America, is remarkable for this semi-barbaric mode of life. A few species grow parasitically as long as they can, and, when sufficient supplies for their full needs are not forthcoming—when they require more nourishment than the heavily-taxed tree they have laid under tribute can afford them—they will send out long shoots to the ground, which take root there, and grow into an actual stem. In other words, like similar units who hang on human society, they only work when they are obliged to do so. They prefer that others should work for them, and would remain permanent parasites if it were possible.

From all of these statements it is evident that the climbing habit, gradually developed by weak-stemmed, ambitious, and crafty plants, so that they can reach as high as bulky-trunked forest-trees, may begin by simply making use of the latter as convenient ladders, and end by strangling them.

Our European climbing plants seldom develop such a murderous tendency. The Honeysuckle will sometimes twist itself so tightly around the slender stems of shrubs it climbs by as to leave deep spiral indentations in their bark, as witness the walking-sticks selected by the curious for the sake of this peculiar appearance. Occasionally, therefore, even this harmless shrub may embrace its ally more tightly than is good for it—may prevent the abundant rise of sap, and the deposition of additional woody tissue. The Lichens which sometimes grow so abundantly on Apple-trees and shrubs, must, according to the late Dr. Lauder Lindsay, abstract all their mineral salts from the plants to which they are attached, for they have no other source to derive them from. Hence the necessity for keeping the Apple-trees "clean," in our cider-growing districts. We can easily understand how this habit might be increased under those intenser habits of growth and competition which would ensue, for instance, if England were once more to enjoy the greater warmth which prevailed here so recently as during the Miocene Period.

But, although our native flora is innocent of producing vegetable Thugs, like those described by tropical travellers, it has a criminal population of its own, even better entitled to the name of "parasites." By this term I mean animals and plants which live wholly or in part upon the *living tissues of other species*. In all these instances the parasite either lives within the tissues of the plant it has attacked, or is engrafted upon them, and fuses its own with them.

Our Mistletoe (*Viscum album*) suggests itself as

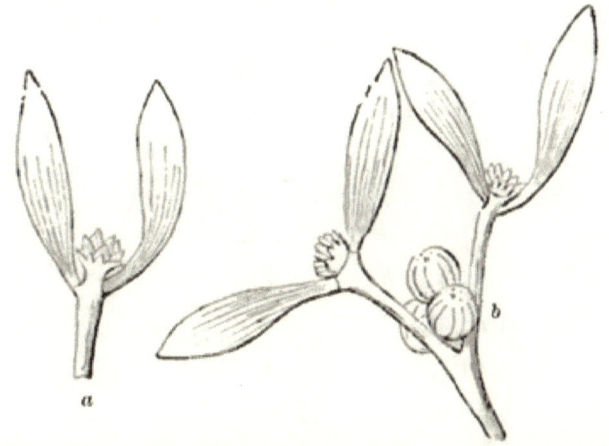

Fig. 82.—Common Mistletoe *Viscum album*); *a*. Flower; *b*. Fruit.

one of the most striking of vegetable parasites. But it is by no means the worst, for it lays the tree on which it grows under only partial tribute, whereas the Dodder and the Broom-rapes (*Orobanchaceæ*) live wholly and entirely upon the food-stuffs and sap stored up and secreted by the plants we find them growing upon. What renders it more disgraceful, on the part of the latter, is the fact that the plants the Broom-rapes attack are naturally small and feeble, such as the Clover, etc.; whereas the Mistletoe and its kind, at any rate, are only found on plants infinitely huger than themselves, and which can well afford to partially sustain them.

I said *partially*, for the Mistletoe bears green leaves, which perform leaf-functions, and abstract carbon from the atmosphere; so that the plant provides itself with starch-stuffs, and only lays the tree on which it grows under contribution for water, and the dissolved minerals the roots have abstracted from the soil and passed upwards into the sap.

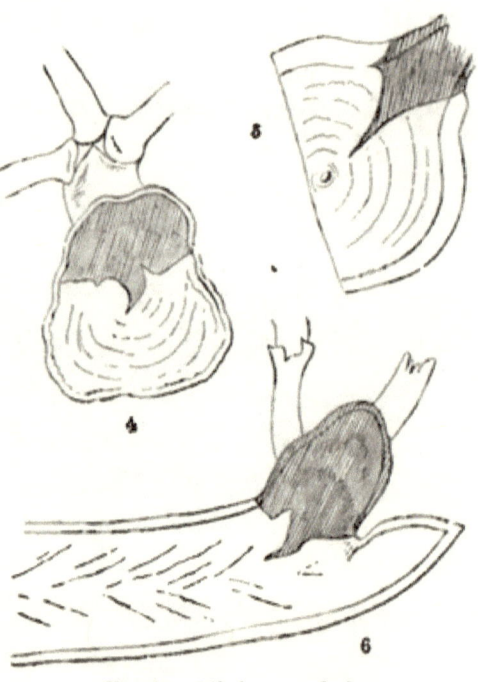

FIG. 83.—Mistletoe grafted.
4. Section across the grain of Apple.
5. Section across the grain of Ash.
6. Section down the grain of Apple.

It is curious, however, to notice the *arboreal* distribution of our common Mistletoe. The tree on which it most *rarely* occurs is the Oak—so rarely that the ancient Druids are said to have regarded its occurrence on it in the light of a great religious event. No other cause can be assigned for the rarity with which the Mistletoe takes up its quarters on the Oak than that suggested by Dr. Carpenter forty years ago, that the latter is rich in *tannin*. Some small portion of this substance must be contained in the sap, and the Mistletoe may not like it, and prefer lodgings where there is less. But, any way, the Oak is the gainer, and can afford to be slighted!

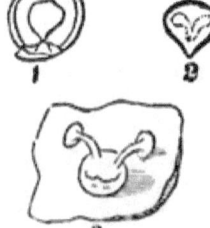

FIG. 84.
1. Vertical section of the fruit of Mistletoe.
2. Vertical section of the seed.
3. (From Baxter) showing the way in which the radicles extend themselves.

Possibly this class of Parasites began originally by being *Epiphytes* —the name given by botanists to all those plants which, instead of growing upon the ground, attach themselves to the back of trees, after the manner of the fruticose Lichens seen on the ancient Apple-trees of our orchards. The Epiphytes are rare in our latitudes, unless represented by Mosses and Lichens, and a few Ferns like the Polypody, which find occasional shelter and provision in the rotting boles of some aged trees. But in tropical countries Epiphytes are a common feature both on forest and on scant vegetation, and the orders to which the Orchids and Pine-apples belong seem to have taken to this artful habit of life more than any other kind of plants, although members of numerous other orders have adopted it. We cannot peruse any work of travel in tropical regions, especially if the writer knows anything of botany, without finding references to and descriptions of beautiful epiphytal Orchids. Growing attached to the trunks of trees, in elevated situations, they are very conspicuous, and as they have no huge stem to build up, we have seen they devote their whole substance to the important act of flowering, the green stem and roots being usually covered with stomata, for the performance of leaf-functions.

But mere attachment to the bark of trees, like that sought by our Ivy, may eventually result in the adventitious roots (at first developed simply as holdfasts) intercepting and obtaining some of the sap of the tree. There are not wanting botanists who hold that the Ivy occasionally indulges in this habit. If this be true, then it furnishes us with an illustration of an epiphyte being transformed into a parasite.

The type of vegetable parasite represented by the common Mistletoe must have been practising the habit for ages. This is indicated both by its root-structure and its geographical distribution. Its root-fibres obey quite a different law to that which governs the early behaviour of the roots of other plants. Whilst the latter grow downward, those of the Mistletoe grow towards the centre of the branch the plant is parasitic upon, and afterwards completely incorporate themselves with the tissues so thoroughly that it is difficult to detect any strong line of demarcation between the two.

The natural order *Loranthaceæ*, to which the Mistletoe belongs, is peculiarly a tropical one, so that our well-known plant is out of its latitude in England. Many of its tropical brethren differ from it in possessing large and brilliant flowers. But the parasitical habit is even more extreme in some foreign kinds than in our own.

FIG. 85.—Mistletoe of the Oak (*Loranthus Europæus*); *a*. Flower; *b*. Fruit.

Gosse describes the habits of a Jamaican species as follows: "What interests me most in this place is a flourishing Mistletoe, or God-bush as the negroes call it. It is growing on a Sour-sop (*Anona muricata*), a tree which it principally affects, overspreading every branch, and effectually, though gradually, killing its supporter. The seeds are viscous, and are to be seen sticking on the leaves and twigs, as well as on the trunk ; in every instance rooting and shooting where they adhere; so that hundreds, perhaps I might say thousands, of young plants, in various stages of forwardness, may be seen on this Sour-sop, springing up from the surfaces of the leaves, three or four on one leaf, and that on both the inferior and superior faces. This I take to be a somewhat unusual phenomenon." From this description it would appear as if the Jamaican Mistletoe had advanced further in parasitism, or vegetable robbery, than our well-known English species.

Again, Sir Charles Bunbury (*Botanical Fragments*) alludes to a Cape species in a manner which shows that it does not differ very much either in structure or habits from our indigenous kinds. "When I visited Uitenhage, on my way back to Cape Town, in the month of June, a beautiful *Loranthus* (*L. glaucus*) was in blossom on the branches of the Acacia, on which it grows parasitically, exactly as our Mistletoe does on European trees. Its flowers are somewhat like those of the Honeysuckle in shape, and of a most vivid orange-scarlet colour. *Loranthus* is a very large genus of parasitical shrubs, almost entirely tropical or sub-tropical; but with one solitary species (*Loranthus Europæus*) growing on Oaks in the south of Europe, and believed by some to have been the original object of Druidical homage—the original 'Mistletoe of the Oak.' *Viscum*, to which our Mistletoe belongs, is a genus of comparatively few species, but with a much wider range than *Loranthus*, occurring in most countries of the temperate as well as torrid zones. I had gathered a *Viscum* at Gongo Soco in Brazil; and I found two in the Cape Colony, one of them remarkable for having no apparent leaves. *Loranthus glaucus* was the only species of that genus I met with at the Cape. I did not see any of those small-flowered kinds which are so ruinous to the Orange and Coffee-trees in Brazil."

This reference to the "ruin" caused by such parasitic plants fully warrants us in regarding them as robbers and even murderers of their own vegetable kind. The Brazilian leafless species of *Viscum* above referred to shows plainly that it has degenerated much further in this parasitical direction than our English Mistletoe, which at least gets half its own food by means of its leaves.

Myzodendron is another genus of woody parasites belonging to the same order as the Mistletoe, which has elected to prey upon the Beeches of Terra del Fuego and Antarctic America. Even the viscid fruits of our favourite English parasite are not better adapted to be gummed to the trees they intend to germinate upon, than those of the *Myzodendron* are for fastening upon the Beech-bark. Each seed of the latter is furnished for this purpose with three long, feathery, viscid bristles, by means of which they cling to the smooth surface. The same arrangement also causes them to adhere to the plumage of the birds sheltering in the trees, which thus unconsciously disseminate them. As soon as the radicle of the *Myzodendron* sprouts, it drives its way through the bark to the growing "cambium-layer" beneath, where it connects itself organically like a graft, and the plant is thereafter supported by the Beech foster-mother, or rather involuntary tax-payer.

Professor Mosely (*Notes by a Naturalist on the "Challenger"*) mentions a species of leafless Mistletoe on the slopes of the Andes, which grows on a leafless Cactus (*Loranthus aphyllus* on *Cereus Quisco*). Such an occurrence presents to a botanist a condensed history of vegetable modification and abortion seldom met with. What a marvellous number of changes must have taken place before this particular and specialised act of parasitism could be brought about! Mosely tells us the parasite is extremely abundant, growing on nearly all the *Cereus* trees; and that it is very conspicuous, because its short stems are of a bright pink colour.

Certain Australian species of Mistletoe bear leaves which can hardly be distinguished from those of the trees on which they live parasitically. No reason is yet known for this singular act of mimicry; but there can be no doubt whatever the simulation has not taken place without a purpose.

The most ingenious of our native vegetable robbers, however, are the Dodders (*Cuscuta*), of which we have several species. There is a refinement about them not indulged in by parasites generally; they are very particular as to the kind of plants they attack. They can only subsist, in fact, upon the sap of certain species, and this, therefore, restricts their parasitism; and consequently the most abundant and widely dispersed of them is *Cuscuta Europæa*, which is least particular, and attacks Thistles, Oats, and, in short, any plants that are crowded together.

FIG. 86.—The Dodder (*Cuscuta epithymum*).

Clover and Gorse are those most preferred by other species; and one, *Cuscuta Trifolii*, confines itself almost entirely to Cloverfields; and another, *Cuscuta Epilinum*, to Flaxfields. It is more than probable, however, that all these so-called species have been differentiated within very recent times; and certainly their specific distinctions are of a very slender kind.

When the seeds of the Dodder drop into the soil they soon germinate, and the little delicate thread-like embryo plant makes its appearance above ground, bearing its seed-covering like a protective cap at its apex. It then looks or feels about for its victim, and dies down in a few days if it cannot find one. The seeds have no cotyledons, like those possessed by the acorn and bean, but they are well provided for instead with a store of albumen (*endosperm*), on which the minute embryo subsists until it is fortunate enough to meet with its prey. As soon as the latter is found, the wire-like stem takes one or two coils around the victim, and develops a series of sucker-like aerial roots which penetrate into its tissues to the upflowing sap. Having its expenditure thus abundantly supplied at no cost to itself, the Dodder grows apace; its red wire-like stems crawl snakewise in and out of the most complicated of host-plants; even the Gorse, in spite of its stiff, prickly leaves, being frequently investured within and without by the triumphant Dodder. The latter is a true Liane, on a diminutive scale, but possessing what none of its tropical representatives have managed to evolve—an elaborate blood-sucking machinery, so that its victim has to support it both mechanically and vitally!

As soon as the young Dodder plant finds it has got a good hold of the proper prey, and has intercepted the sap supplies after the manner just described, it lets go its hold upon the soil where it germinated, but not before. No fewer than four genera of Dodder parasites have been enumerated by botanists, including fifty species, all of which obtain their livelihood after the same nefarious fashion. So long and so successfully have the Dodders practised thievery that most if not all the above species would become extinct if they were obliged suddenly to reform and alter their habits of life.

When we remember that our pretty Convolvuluses belong to the same order as the Dodders (*Convolvulaceæ*), we get a glimpse of how this confirmed and highly-developed parasitic mode of life may have originated. Most of the Convolvuluses have either a twining or a creeping habit; and some of them twine so successfully that they go by the popular name of Bindweeds. So far, therefore, they have the same initial habit as the Dodder. The latter, having commenced life thus as a hanger-on, proceeds to its next stage as a parasite. Its leaves are gradually aborted, until ultimately they are not even produced in a rudimentary state; and thus the Dodder has grown leafless. Its dense clusters of flowers are those of the ordinary Convolvulus on a diminutive scale, and they produce numerous seeds, all at the expense of the victimised host-plant. The Dodders must have been at their trade an immense period of time, for nearly all of their original organs have disappeared, and their parasitic habits are about as highly developed as any in the world.

As if others had copied the evil but successful mode of life thus set them, Sir Charles Bunbury remarks that a "parasitic plant, common in the neighbourhood of Cape Town, both on the hills and on the flats, is the *Cassytha filiformis*, which has exactly the general form and appearance of our Dodders; but, strange to say, the structure of its fruit and flowers obliges us to class it with the *Laurel* family, to which it has otherwise not the slightest resemblance. It has no leaves, and its stems are lax, slender, yellowish threads, which ramble far and wide over the bushes, and lace them with a strange kind of web; the flowers greenish-white, very small and inconspicuous. If it really descended from the same original stock with the magnificent Laurel-trees of tropical forests, the time when it branched off from them must have been inconceivably remote."

But we have a tolerably large tribe of vegetable robbers which practise their craft underground instead of upon trees and shrubs, and whose rootlets feel out for and seize on the roots of other plants. Their flower-stems then rise above the ground, looking as if théy had been developed in a fair and honest manner. Their brown, sometimes purplish, flowers are of high organisation, but their leaves are reduced to mere brown *scales*—so long is it since they performed any legal and honest leaf-functions! The most notorious of these parasites are the Broom-rapes (*Orobanchaceæ*), popularly so called because there are few of our ancient heaths and commons where this class of plants does not grow upon the roots of the Broom and even of the Gorse. We may also see them in every Cloverfield, where their comparatively huge stems are brutally forcing the

Fig. 87.—Broom-rape (*Orobanche rapum*). The scales on the stem are useless and aborted leaves.

impoverished Clover roots to support them gratuitously. Thus they flourish in the midst of plenty, without putting forth any effort of their own whatever, either in the way of assimilating mineral salts from the soil, or of abstracting carbon from the atmosphere; whilst their general bitterness and astringency protects them from the grazing animals attracted by the Clover and other edible plants: and so we see them distributed all over our fields, erect and triumphant, the very personification of successful knavery!

All our British members of the natural order *Orobanchaceæ* are criminals in this respect, and there are eight species of them, parasitically growing on the roots of Broom, Gorse, Knapweed, Clover, Goose-grass, Wild Thyme, Yarrow, Hemp, etc. The roots of other plants are subject to the predatory assaults of similar foes, such as the Toothwort (*Lathræa squamaria*), to be met with in April on the roots of Alders by stream sides, as well as on those of the Hazel in damp spring woods. The nearly-allied order, *Scrophulariaceæ*, contains parasites in what might be almost called "every stage of parasitic degeneration," such as those which only occasionally resort to it; others which adopt the habit whenever they can; and a few which cannot live unless they are allowed to prey upon some other plant. Among the latter is the *Harveya* of the Cape, which is wholly parasitic upon the roots of the Heaths, and looks to all the world like a genuine Broom-rape, but the structure of its seed-vessel proves it to belong to the *Scrophulariaceæ*.

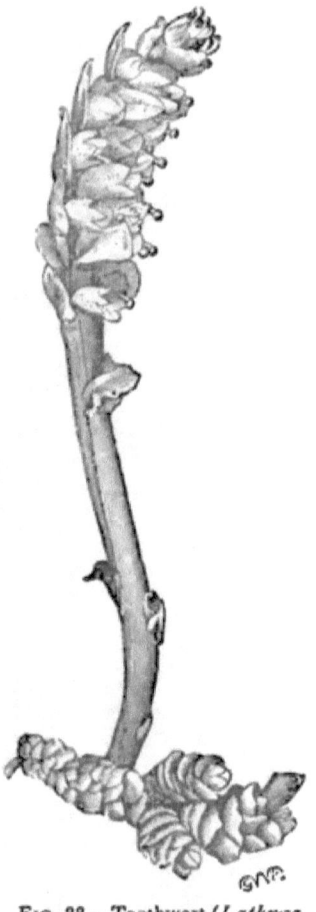

FIG. 88.—Toothwort (*Lathræa squamaria*).

The Yellow-rattle (*Rhinanthus crista-galli*), so abundant in our summer meadows, and the Eyebright (*Euphrasia officinalis*) of our dry pastures and green hillsides, are guilty of partial parasitism. Cow-wheat (*Melampyrum*), Toad-flax (*Thesium*), and others practise it, perhaps in a less degree.

In each and all of these instances of vegetable robbery and murder the plants engaged in it have degenerated from higher, nay, usually from the very highest-developed forms. None have *advanced* towards these practices, although the long indulgence in them has in many cases produced contrivances and adaptations as striking as those put forth for more honest purposes. The only cause assignable for the parasitism of these higher plants is poverty —which forces them to choose between parasitism or extinction !

What is to be said of that host (whose name is Legion) of lower organised plants which never appear to have adopted any other mode of life than that of robbing and murdering, and generally preying upon, plants much higher in vegetable rank than themselves? To such belong our parasitic Fungi—the dread of agriculturists and horticulturists, from the extent of their ravages, their virulence, and their unsurpassed cunning in evading detection and interference. No wonder ! In the first place there is reason to believe this class of plants never lived in any other way than by plundering more highly-organised plants of their vegetable wealth; in the

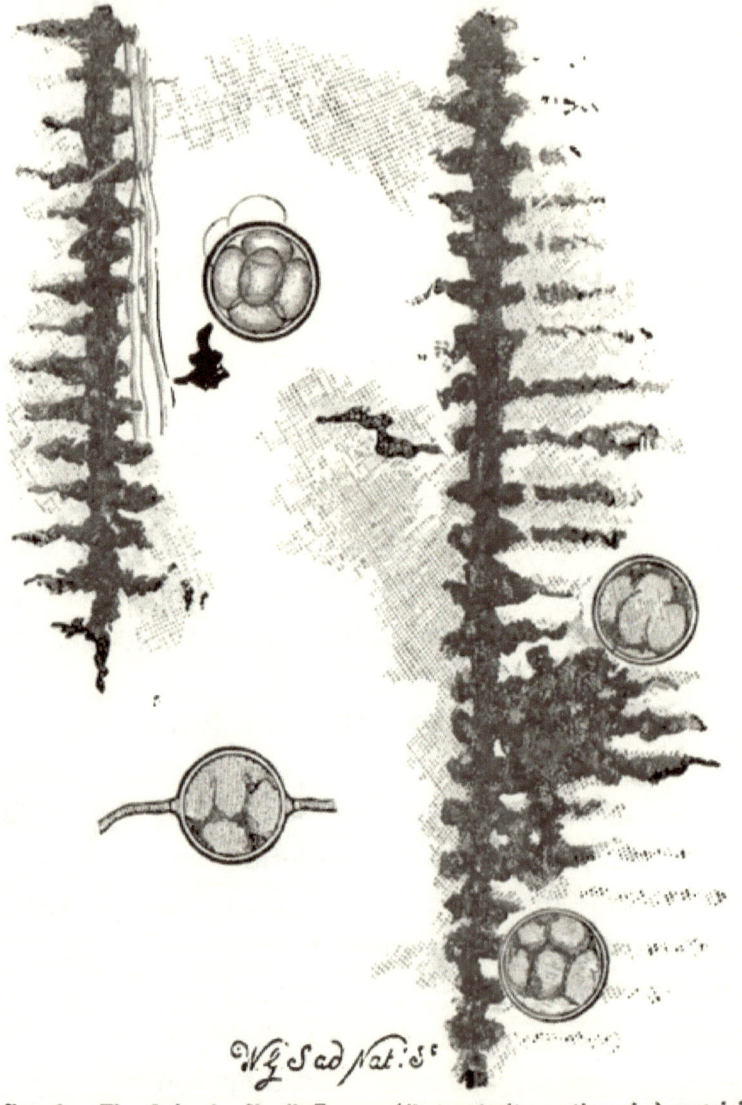

FIG. 89.—The fruit of a Fossil Fungus (*Peronosporites antiquarius*) containing zoospores *in situ*, as seen amongst the scalariform vessels of a Lepidodendron from the Coal measures (enlarged 400 diam.)

second, they have been practising the habit for

untold millions of years, for we find the fossil *Lepidodendra*, of the Carboniferous Period, riddled and perforated by the minute interpenetrations of a parasitical Fungus not distantly related to our too well-known Potato-disease genus (*Peronospora*).

Dr. M. C. Cooke, the distinguished fungologist (*Fungi, their Nature, Influence, and Rises*), says of these ubiquities and geologically hereditary vegetable robbers: " In the other section of the *Coniomycetes* (Dust-funguses) the species are parasitic upon, and destructive to, *living* plants, very seldom being found on really *dead* substances, and even in such rare cases undoubtedly developed during the life of the tissues. Mostly the ultimate stage of these parasites is exhibited in the ruptured cuticle, and the dispersion of the dust-like spores; but in *Tilletia caries*, *Thecaphora hyalina*, and *Puccinia incarcerata*, they remain enclosed within the fruit of the foster-plant. The different genera exhibit in some instances a liking for plants of certain orders on which to develop themselves, *Peridermium* attacks the *Coniferæ*; *Gymnosporangium* and *Podisoma* the different species of Juniper; *Melampsora* chiefly the leaves of deciduous trees; *Rœstelia* attaches itself to pomaceous (apple-fruited) trees; whilst *Graphiola* affects the *Palmaceæ*; and *Endophyllum* the succulent leaves of the house Leek. In *Æcidium* a few orders seem to be more liable to attack than others, as the *Compositæ, Ranunculaceæ, Leguminosæ, Labiatæ,* etc.; whilst others, as the *Graminaceæ, Ericaceæ, Malvaceæ, Cruciferæ*, etc., are exempt. There are, nevertheless, very few orders of phanerogamous plants in which some one or more species, belonging to this section of *Coniomycetes*, may not be found; and the same foster-plant will occasionally nurture several forms. Recent investigations tend to confirm the distinct specific characters of the species found on different plants, and to prove that the parasite of one host will not vegetate upon another however closely allied."

Dr. Cooke very generously and euphoniously speaks of the particular kinds of plants whose life-histories he has studied with so

FIG. 90.—*d*, two uredo spores of the *Puccinia graminis* germinating upon the cuticle of a wheat leaf; the germ-tube of the lower spore has just entered a stomatum; in the upper spore the process is more advanced.

much diligence and success, in a tender and even affectionate manner. The victims are called "hosts" or "foster-plants;" but it is not difficult to see that all they get by their generous hospitality in entertaining such pests is robbery and murder, and that they are only "foster-plants" in the sense that the bird was foster-mother to the serpent's eggs which ancient fable says were placed under her to hatch, and whose successful result was to issue forth and destroy the pseudo-parent!

Our mildews, rusts, smuts, potato, vine, hollyhock, and other diseases, our pea-blights, rose-blights, etc., only too surely proclaim the successful depredations of these vegetable barbarians, strong in their numbers and ferocity, like those hordes of Goths and Huns who burst like a flood upon and overcame the high civilisation of ancient Rome.

CHAPTER XIII.

"TURNING THE TABLES."

THE crimes of "robbery and murder," as illustrated by plant-life in the preceding chapter, were entirely confined to the vegetable kingdom. Nothing was there said of the fact that many plants are so virulently poisonous as to cause death to animals—that peculiarity has been already considered as protective. Nor have the habits of such wonderful and highly-elaborated pieces of vegetable mechanism as the Venus' Fly-trap (*Dionæa muscipula*), the Sundews (*Droseræ*), the Pitcher-plants (*Sarracenia, Nepenthes, Cephalotus*, etc.), or of our humbler Butterworts (*Pinguiculæ*), been mentioned particularly as cunningly devised machinery for insect-assassination.

But it now remains for me to relate the strange means by which certain kinds of plants have been able "to turn the tables" on their ancient and hereditary foes the insects. The latter serve the plants many a good turn, it is true—as when they are engaged in fertilising their flowers; but this is performed by a few only out of the innumerable hosts of insects; and even the larvæ of the former make the leaves of the plants, whose flowers are benefited by their parents, pay fine and toll for the service.

It may be that the remarkable habit of catching flies, and afterwards of making a meal of them, has been developed by adversity. I pointed out six years ago, in *Flowers: their Origin, etc.*, that "The marsh-loving habits of most of these plants, both British and foreign, show that they usually grow in places where their roots can absorb but little if any nitrogenous material. This duty is therefore thrown upon other parts of the plant, some of which are normally in the condition that the spongioles of the roots are; so that when decomposing animal matter comes into contact with them, they can absorb it." To the difficulty which marsh-plants have of getting nitrogen might be added that of obtaining potash and other salts, with which, however, captured insects would provide them. Our English Sundews (*Droseræ*) may often be seen growing in myriads on the surface of Sphagnum bogs, their slender roots merely anchoring them to their places, and perhaps providing them with water; whilst the marvellously altered and adapted leaves not only obtain carbon from the atmosphere, but artfully contrive to get all the nitrogen, potash, etc., the whole plant requires from the capture of insects!

Such plants have been called "insectivorous" and also "carnivorous." Darwin has demonstrated that those fed with animal food produced more flowers and seeds than those left alone. There can be no doubt, therefore, that this habit is a highly-specialised one, and that it has been acquired, for it is practised in varying degrees — from occasional indulgence to absolute necessity. And what is very striking is the manner with which fly-catching and digesting has been adopted by plants of various orders, in several ways, all over the world. Some of these set regular traps, like the Sundews, Venus' Fly-catcher, etc.; others, like our Butterworts, grease their leaves and curl up their edges. Many kinds in America grow specialised pitcher-shaped leaves, and smear their upper surfaces with a honey-like secretion, whose sweetness is intensified within, alluring the unsuspecting flies on until they cannot return. They then accumulate in a seething, filthy, half-dead, half-living mass, within the interior of this diabolically contrived trap—a mere manure-heap for the benefit of the plant, whose seeds will be all the richer in albumen from the transformation of the organic matter of the insects by the process of vegetable digestion or assimilation.

When members of the same genus of plants are found in various parts of the world practising the same habits, with such a remarkably similar mechanism, there is only one explanation I know of

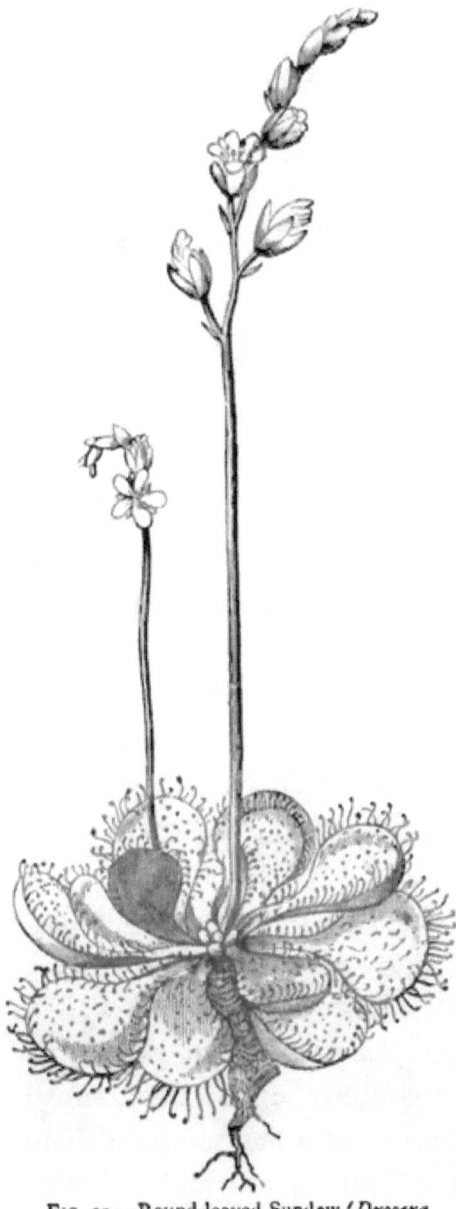

Fig. 91.—Round-leaved Sundew (*Drosera rotundifolia*).

which can be offered—the genus must be a very ancient one, and its geographical distribution must have commenced before the present configurations of sea and land existed.

Take the Sundews (*Droseræ*), of which we have three British species, one (*rotundifolia*) being very common. The leaves are covered with red hairs, developed into glands, whose summits secrete a clear drop of what looks like dew (whence the popular name), but which is in reality such a viscid substance that it may be drawn out to the tenuity of a spider's line. Moreover, as the reader will see by referring to Darwin's sagacious experiments, these red "tentacles" are sensitive; nay, they even know what is good to eat and what is worthless! If a fragment of meat or boiled egg is placed on the leaf, the tentacles gradually bend over, the edge of the leaf by curling assists them, the viscid secretion is poured out in greater quantity as if stimulated, until presently the morsel is completely enveloped, and the process of digestion or assimilation commences. When it is over the leaf flattens out to its old position, the tentacles regain their erectness, their tips become globular with viscid dew, and thus the trap is once more set.

FIG. 92.—Leaf of *Drosera anglica*.   FIG. 93.—Leaf of *Drosera obovata*.

But in a state of nature the leaves of the Sundew know nothing of chopped beef-steak and fragments of hard-boiled egg. They recognise insect visits, however, for they are an instinctive tradition to all genuine flowering plants. Therefore, for the particles mentioned in the above experiment, substitute flies; and any observer who has seen the helpless, frantic struggles of insects wishful to allay their thirst, who have been attracted by the glistening drops of seeming moisture secreted by the tentacles (assisted by the appeal made to their generally correct sense of colour by the bright red of the organs) into alighting upon the leaves—almost feel that nothing less than an Act of Parliament ought to be introduced to put an end to such insidious cruelty!

These tentacles are totally unaffected if particles of sand or any other inorganic substance is placed upon the leaf, and they never notice or respond even to drops of water. This fact raises their character for instinctive sagacity in knowing what is harmful or unnecessary to them, as well as what is good in the shape of animal food. To the pattering of raindrops the tentacles must be well accustomed, and occasionally the wind will carry grains of sand and earth to them, and cover them with dust. In this way the tentacles have gained practical experience as to what is good for them and what is not.

These delicate, highly-organised glands or tentacles are merely *hairs* (common to the leaves of most plants) which have been gradually differentiated into their present structures and functions. The hairs on the stems and leaves of plants are capable of wonderful modification. Sometimes they are stiffened into defensive prickles, as in the Gooseberry; at others they are converted into stings as in the Nettle. In the Nottingham Catch-fly and other plants they secrete a viscid fluid to prevent ants and other creeping insects from climbing the flower-stem.

"TURNING THE TABLES." 263

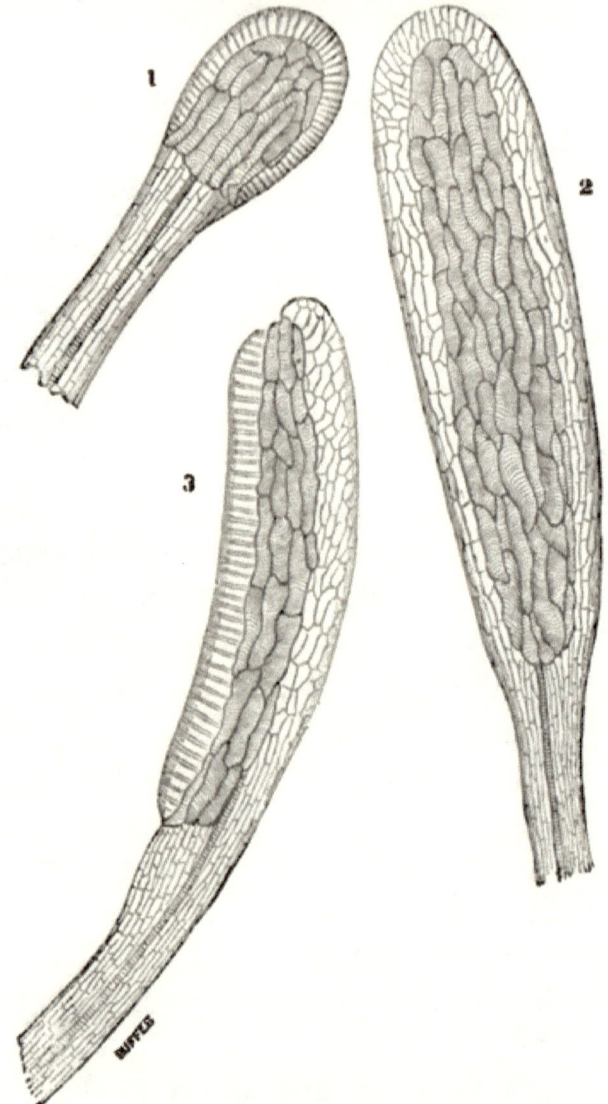

Fig. 94.—Glands of *Drosera rotundifolia*.
1. Oval gland of *D. rotundifolia*, × 130.
2. Cylindrical gland (front view) of *D. rotundifolia*, × 130.
3. Ditto (side view) of *D. rotundifolia*, × 130.

It seems but a step farther in advance of this last

function for such modified hairs to be converted into glands which shall secrete a viscid fluid capable of digesting albuminous substances, like the tentacles of the Sundews, and I strongly suspect that the Henbane (*Hyoscyamus niger*) is now in such a transition state, for its sticky stem is always covered with dead and dying flies.

The Sundew family has a marvellously wide geographical distribution, considering the peculiar habits of its members. About one hundred species are known, belonging to several genera, all of which partake in the Sundew mode of life. Nearly all affect the same physical conditions. They are essentially marsh-loving plants, and in such places are found throughout Europe, India, China, the Cape of Good Hope, Madagascar, North and South America, and Australia—practically a cosmopolitan distribution. How singularly their habits and structure coincide in species inhabiting countries so far apart, and what a remarkable dispersion the same species sometimes has, will be seen from the following remarks in Kingsley's *At Last*, of a specimen he found in the Savannah of Aripo: "My kind guide put into my hand, with something of an air of triumph, a little plant, which was, there was no denying it, none other than the long-leaved Sundew (*Drosera longifolia*), with its clammy-haired paws full of dead flies, just as they would have been in any bog in Devonshire or in Hampshire, in Wales or in Scotland. But how came it here? And more, how has it spread, not only over the whole of northern Europe, Canada, and the United States, but even as far south as Brazil?"

Practically, the Sundews belong to the southern rather than the northern hemispheres, and they appear to have been driven thence across the line to our latitudes, perhaps during the great Southern Glacial Period which drove many other southern plants to northerly climates. "On the Organ mountains of Brazil, both Arctic and Antarctic plants are found commingled in strange brotherhood, eloquently testifying to the alternate glaciation of the northern and southern hemispheres, which has thus unexpectedly brought them into company." No fewer than forty-one species of Sundew have been found in Australia alone. Bunbury thus describes the South African kinds: "Of the curious and interesting genus *Drosera*, there are seven species found near Cape Town, all as remarkable as those of Europe for their delicate clothing of gland-tipped hairs, giving them that peculiar *dewy* appearance from whence they have the names of *Drosera* and *Sundew*. *Drosera capensis*, which grows in wet and boggy spots on the sides of Table Mountain, is similar in form and in the arrangement of its flowers to our *anglica*, but much larger in all its parts, and with beautiful bright purple flowers. The most peculiar species, and at the same time the most common about Cape Town, is the *Drosera cistifolia*, which, unlike all the others that I have seen of this genus, is not at all confined to bogs, but grows on the hard rough ground sloping down from the mountains, as well as on the sandy flats. I must add, however, that it appears only during the wet season."

Even more remarkable in the specialised struc-

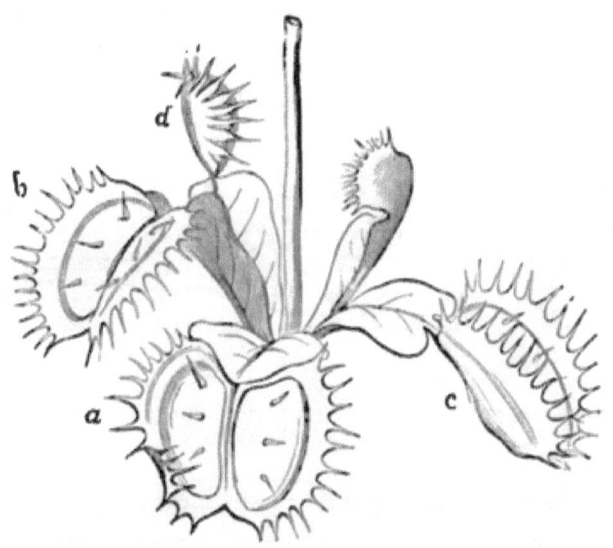

Fig. 95.—Venus' Fly-trap (leaf open at *a*; partially closed at *b*; and almost closed at *c*).

ture of its leaves for fly-catching is the well-known Venus' Fly-trap (*Dionæa muscipula*). The foot-stalks of the leaves are flattened out, covered with stomata, and perform the functions of true leaves, so that they leave the latter free to be adapted to other duties. This remarkable plant has its natural home in the damp and swampy places of North Carolina.

The edges of the true leaves are set all round with stiff bristles, which interlock like the fingers of the clasped hands when shut, and if anything happens to be included within them it is held prisoner there. If it should be a particle of meat or a fly it is retained, and subsequently digested by a fluid which oozes forth, and the organic matter is thus assimilated. When the blades of the leaf are opened, three stiffish hairs may be seen on each half, standing upright like sentinels; these are highly sensitive (Fig.

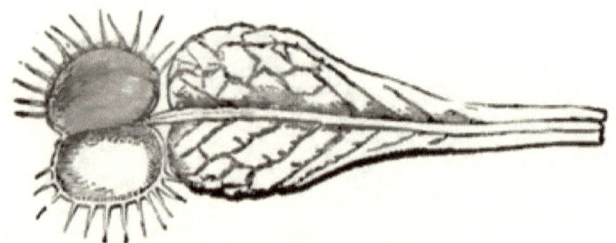

FIG. 96.—Leaf of Venus' Fly-trap; lower part is a phyllode or flattened leaf-stalk, doing ordinary leaf-duty.

95 *a*). A touch with a hair is sufficient to enable them to commence a series of internal protoplasmic changes of such rapidity that the blades close immediately like a mouse-trap. If an insect alights upon the leaf, and touches these sensitive hairs, it is then captured, strangled, and digested. But it may walk about with safety so long as no part of its body comes into contact with the telegraph instruments, as the hairs might be called. The mucilaginous fluid secreted by the leaves is of an acid nature, and does not take long in dissolving out of the imprisoned dead insects

all the substances the leaf requires. Then the leaf slowly opens and expands, the hairs are erect and on duty, and the trap is ready for another victim.

Darwin, in his *Insectivorous Plants*, narrates in his quiet but effective manner the numerous experiments made upon all kinds of flesh-eating plants, and with the Sundews and Venus' Fly-trap in particular, and to that notable book I refer all readers who now hear for the first time of plants that have turned the tables on their enemies, and devour them, just as the most savage races of mankind do their prisoners of war.

Moreover, a lady naturalist, Mrs. Mary Treat of New Jersey, found that these insectivorous plants suffered from *indigestion* when they attempted to consume too many flies at a sitting, and that some even died from over-eating! Her remarks are as follows: "Several leaves caught successively three insects each, but most of them were not able to digest the third fly, but *died in the attempt.* Five leaves, however, digested each three flies, and closed over the fourth, but *died soon after* the fourth capture!"

Dr. Burdon Sanderson has demonstrated that the contraction and contractibility in the tissues of the leaves of Venus' Fly-trap are identical in their character with those which take place when *muscular contraction* occurs in the muscles of mammalia. Moreover, he has further shown there is a correspondence between the electric phenomena which accompany muscular contraction and those which are associated with the closing of the leaves of the Venus' Fly-trap.

The two groups of plants just described capture their insect prey after two different methods, each of which has been separately evolved as a vegetable policy. We now approach a third—that adopted by all those plants (belonging to several orders widely separated, and also geographically isolated), some of whose leaves have been structurally modified, not into fly-traps, but into *pitchers*. The very device we see in shop-windows, of attracting the hosts of house-flies, by means of a sweetened liquor to enter glass vessels from which they cannot emerge, but accumulate and die and seethe there, was adopted by the North American and other "Pitcher-plants" ages ago! North America is especially distinguished by the production of this sort of fly-trap, for we find no fewer than eight species represented there, all of which are marsh-loving, or bog-haunting plants. The *Sarracenias* have been long noted for the remarkable alteration in the shapes of their leaves. Some of the "pitchers," formed by their edges growing together (in a way not uncommon as a "monstrosity" in cultivated plants, such as the Cabbage (Fig. 97), are three feet in height, as those of *Sarracenia flava*, for instance. Most of the "pitchers" have lids or covers, which do not fit close, but seem rather intended to prevent the dust settling within, or the direct heat of the sun from evaporating the fluid contents. Being highly coloured, they doubtless serve also to attract the insect victims, and induce them to alight upon the plant.

Nothing could be more effectively devised to ensure success than the shape and secretion of these "Pitcher-plants." They rival even those temptations to young men, which, however attractive at first, become more and more seductive and enchanting, until the fatal step is taken, and recovery is impossible! Thus, the lips of

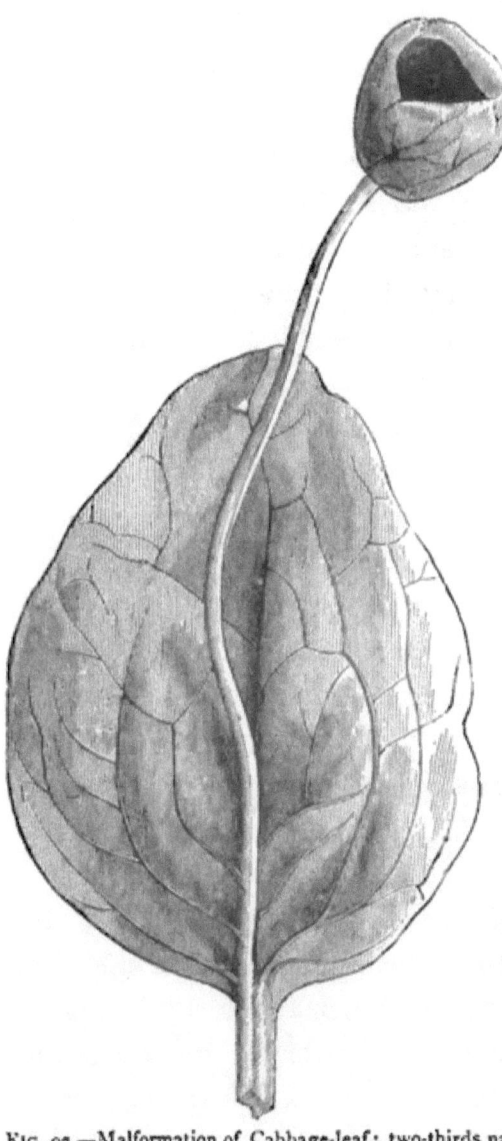

FIG. 97.—Malformation of Cabbage-leaf: two-thirds natural size, showing how a natural malformation of leaf may form a pitcher or cup.

these "pitchers" secrete a honeyed fluid, delightful to the hungry and thirsty insects which have alighted on the attractively-coloured and patterned plant. This they greedily devour, following it up, or rather down, the "pitcher," where it becomes sweeter and sweeter. The hairs and other processes bend inwards and downwards and offer no resistance to the thoughtless prey, but they absolutely bar return, and the deluded flies drop one by one into the horrible dungeon at the bottom of the tube! Professor Asa Gray, the distinguished American botanist, has thoroughly studied both the structure and habits of these plants, and he thus describes them and their victims: "After turning back the lids of most of the leaves, the flies would enter as before, a few alighting on the honeyed border of the wing, and walking upward, sipping as they went to the mouth, and entering at the cleft of the lower lip; others would alight on the top of the lid, and then walk under the roof, feeding there; but most, it seemed to me, preferred to alight just at the commissure of the lips, and either enter the tube immediately there, feeding downward upon the honey pastures, or would linger at the trunk, sipping along the whole edge of the lower lip, and eventually near the cleft. After entering (which they generally do with great caution and circumspection), they begin again to feed, but their foothold for some reason or other, seems insecure, and they occasionally slip, as it appears to me, upon this exquisitely soft and velvety declining substance. The nectar is not exuded or smeared over the whole of this surface, but seems disposed in separate little drops. I have seen them regain their foothold after slipping, and continue to sip, but always moving slowly and with apparent caution, as if aware that they were treading on dangerous ground. After sipping their fill they frequently remain motionless, as if satiated with delight, and, in the usual self-congratulatory manner of flies, proceed to rub their legs together, but in reality, I suppose, to clean them. It is then they betake themselves to flight, strike themselves against the opposite sides of the prison-house, either upwards or downwards, generally the former. Obtaining no perch or foothold, they rebound off from this velvety, microscopic *chevaux de frise*, which lines the inner surface still lower, until by a series of zigzag, but generally downward falling flights, they finally reach the coarser and more bristly pubescence of the lower chamber, where, entangled somewhat, they struggle frantically (but by no means drunk or stupified), and eventually slide into the pool of death, where, once becoming slimed and saturated with these Lethean waters, they cease from their labours. After continued asphyxia they die, and after maceration they add to the vigour and sustenance of the plant. This seems to be the true use of the limpid fluid, for it does not seem to be at all necessary to the killing of the insects (although it does possess that power), the conformation of the funnel of the fly-trap is sufficient to destroy them. They only die the sooner, and the sooner become liquid manure."

Spenser could not have planned a more subtle or cunning story of temptation, deceit, and final ruin! For "pitchers" and "flies" we have only to substitute the various dissipations of "fast life," and the characters who indulge in them—and both fact and moral hold equally true!

Dr. Riley, the distinguished American entomologist, in an article contributed to *Science Gossip* in 1874, sums up the evidence in favour of the *Sarracenias* being insectivorous plants as follows: "There is no reason to doubt, but every reason to believe, that *Sarracenia* is a truly insectivorous plant, and that by its secretions and structure it is eminently fitted to capture its prey.

FIG. 98.—Leaves of the Sarracenia, or Side-saddle Plant, one of the Pitcher-plants.

T

"That those insects most easily digested, and most useful to the plant, are principally ants and small flies, which are lured to their graves by the honeyed paths; and that most of the larger insects which are not attracted by sweets, get in by accident, and fall victims to the peculiar mechanical structure of the pitcher.

"That the only benefit to the plant is from the liquid manure resulting from the putrescent captured insects."

It seems a strange thing that plants should in this way obtain their own manure, but it must be remembered that in reality manure is one kind of plant-food, and that it matters nothing eventually whether such food is allowed first to become putrescent, as with the *Sarracenias*, or is strangled alive and then digested, as with the Sundews and Fly-traps. The difference is that the latter kind of carnivorous plants have to be at the trouble of digesting their prey, whereas the former are saved that process.

Darlingtonia is another kind of "Pitcher-plant," found growing high up the Californian mountains. Its general appearance, with its mottled cap or hood, and the two-lobed leaves dependent from it, resembles the head of a snake, with tongue protruded. The pitcher is usually found crammed with insects, chiefly moths, in a putrescent state.

Nepenthes is the old-fashioned "Pitcher-plant" of our conservatories, and was so named before the American *Sarracenias* and *Darlingtonias* were imported into this country for the sake of their insectivorous habits. It is abundant in the lands of the Malayan Archipelago. Dr. A. R. Wallace, in his well-known work on that region, speaks as follows of those he saw in Borneo: "The wonderful Pitcher-plants, forming the genus *Nepenthes* of botanists, here reach their greatest development. Every mountain-top abounds with them, running along the ground or climbing over shrubs, and stunted trees; their elegant pitchers hanging in every direction. Some of these are long and slender, resembling in form the beautiful Philippine Lace-sponge (*Euplectella*), which has now become so common; others are broad and short; their colours are green, variously tinted, and mottled with red or purple. The finest yet known were obtained on the summit of Kini-Balou, in north-west Borneo. One of the broad sorts (*Nepenthes rajah*) will hold two quarts of water in its pitcher. Another (*Nepenthes edwardsiana*) has a narrow pitcher 20 inches long, while the plant itself grows to the length of 20 feet."

According to Sir Joseph Hooker, no fewer than thirty species of Nepenthes are known, chiefly from this region and Ceylon, as well as Australia, etc. In all cases the mouth of the Pitcher is furnished with a thickened corrugated rim. This not only strengthens the mouth and keeps it open, but it also secretes a *sweet liquid*, after the manner of the Sarracenias, and for the same purpose. Insects attracted by it crawl farther and farther down, and cannot return, owing to the recurved hooks, hairs, and overhanging rim. These hooks are often strong enough to catch small birds which come to drink at the water usually to be found in the pitcher, and then they suffer the same fate as the flies.

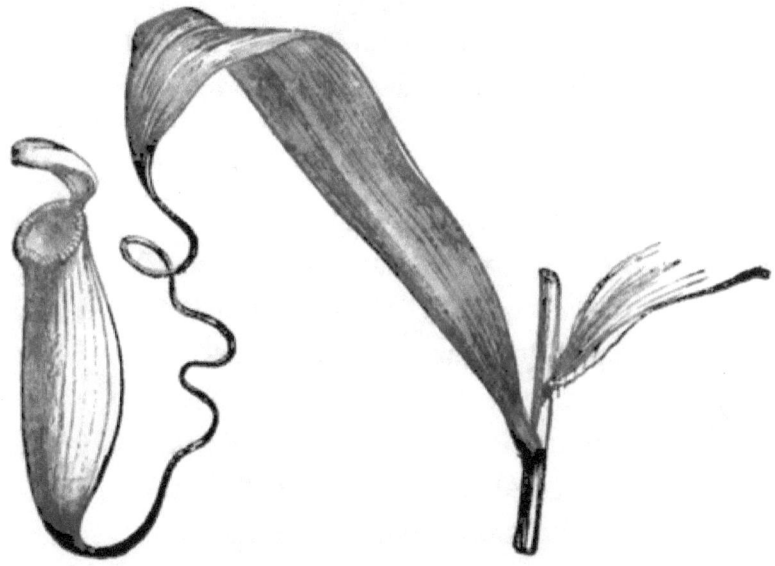

FIG. 99.—Nepenthes, or Pitcher-plant.

The larger and peculiarly-constructed *Nepenthes*, therefore, may be ornithivorous as well as insectivorous. One species, *Nepenthes bicalcaratus*, derives its specific name from the two long pointed hooks which are directed downwards towards the mouth of the pitcher, and these cannot fail to prevent large insects or even small birds from getting out after they have once got in.

Glands, which have been proved to secrete an acid fluid, line the interior of the pitcher and pour their contents into it, so that the water which accumulates there is always acid. Experiments have shown that pieces of fresh raw meat, egg, cartilage, etc., are actually dissolved therein, the cartilage being converted into jelly in the space of three days. Sir Joseph Hooker delivered an address at the Belfast Meeting of the British Association on "Carnivorous Plants," in the course of which, speaking of the experiments which had been made on the *Nepenthes*, he said: "From these observations it would appear probable that a substance acting as pepsine is given off from the inner wall of the pitcher, but chiefly after placing animal matter in the acid fluid."

Dr. M. C. Cooke (*Freaks and Marvels of Plant Life*) thus summarises the general structure of these remarkable vegetable structures: "That they are adapted for the capture and retention of insects; that, at the orifices, are certain attractions, such as the production of a sweet fluid, which would be likely to allure insects into the traps; that the mouth is protected by an overhanging lid, which would prevent the falling in of small and useless objects, but insufficient to obstruct the entrance of living prey; that this lid may also prevent the admission of an excess of external moisture; that the internal structure is extensively glandular, the glands being elevated, but at the same time protected by a hood; that the glands are largest, cover the largest surface, and are least protected at the bottom and the lower third of the inner surface; that these glands secrete a digestive fluid, with an acid reaction; that insects are commonly found at the bottom of the pitchers, where they become disintegrated; that they are probably digested by the excreted fluid, and their soluble nitrogenous matter absorbed and assimilated for the advantage of the plant. . . . From all this, if the summary is a fair one, we are naturally led to conclude that the pitchers of Pitcher-plants are traps to catch animals, as well as stomachs to digest them."

Another well-known but much smaller Pitcher-plant, often to be seen in greenhouses, is *Cephalotus*, from Australia. It is a small plant, with rosette-clustered leaves, from which the stalks of the pretty little pitcher radiate. The general mechanical structure of the latter is wonderfully like that of *Nepenthes*, and the pitchers are provided with lids, probably for the same reason. Insects are caught and digested in pretty much the same way.

It is hardly necessary to point out that these insectivorous habits must have all been acquired, and that the different stages of the acquisition may still be seen in a variety of plants. The most interesting fact in the habit, perhaps, is how plants have approached it from different directions, so to speak, and performed it by different mechanical contrivances.

It is hardly less instructive to notice how plants belonging to widely-separated orders, and having no affinities with each other, to say nothing about their being seperated by great geographical distances and natural barriers, like the *Sarracenias, Nepenthes,* and *Cephalotus,* have hit upon the same general device.

Our English flora is represented by other flesh-feeding or carrion-feeding species. Chief among them is the pretty and conspicuous Butterwort (*Pinguicula vulgaris*), with its violet coloured and shaped flower borne on a tall stem from the midst of a pale-green rosette-shaped cluster of leaves, in wet and boggy places. Those leaves are very greasy, whence the plant's name; and it is their greasiness which prevents small flies that have alighted upon them from getting away. Nay, their very slippings about and struggles only incite the glands to secrete more "butter," and the edges of the leaf curl up to prevent its sliding or trailing its body away. Then again, we have the Bladderworts (*Utricularia*) of our sluggish streams, better known by their spikes of small yellow flowers appearing above the surface, than by the remarkable structure of their submerged "bladders." The latter were formerly believed to exist for the purpose of buoying up the plant, and in part they may still serve that purpose; but Darwin and others have shown that they are genuine insect-traps, constructed on the same mechanical principle as an eel-trap, very easy to enter, but impossible to break out of. Into these, small aquatic insects creep and die. Nearly every "bladder" on a plant is found to contain an insect, generally a minute water-beetle or water-flea, the latter being apparently the chief prey of the plant. The mouth of this "trap" is set with bristles, or stiff plant-hairs, and there are others which point inwards, so that insects easily get in but cannot get out. Species of Bladderworts are distributed throughout Europe and North America, as well as in the West Indies and Brazil; and Mrs. Mary Treat, already mentioned for her experiments on other plants, has watched the habits of a North American species, *Utricularia clandestina*, very closely, and put on record some very interesting facts in connection with its mode of action. After speaking of water-bears, water-fleas, etc., which had been admitted within the "bladders" and kept there, she says: " So these points were settled to my satisfaction—that the animals were entrapped and killed, and slowly macerated.... I found almost every bladder that was well developed contained one animal or more, or their remains, in various stages of digestion.... There was some variation with different bladders as to the time when maceration or digestion began to take place, but usually, on a growing spray, in less than two days after a large larva was captured the fluid contents of the bladder began to assume a cloudy or muddy appearance, and often became so dense that the outline of the animal was lost to view. Nothing yet in the history of carnivorous plants came so near the animal as this. I was forced to the conclusion that these little bladders are in truth like so many stomachs digesting and assimilating animal food."

Charles Kingsley's observant eye did not fail to notice the peculiarity of the West Indian kinds of Bladderworts: "The type of the rushes and grasses was very English; but among them grew, here and there, plants which excited my astonishment; above all, certain Bladderworts, which I had expected to find, but which, when found, were so utterly unlike any English ones, that I did not recognise at first what they were. Our English Bladderworts, as everybody knows, float in stagnant water on tangles of hair-like leaves, something like those of the Water-Ranunculus, but furnished with innumerable tiny bladders; and this raft" (Kingsley held the theory that the "bladders" were buoys) "supports the little scape of yellow snapdragon-like flowers. There are in Trinidad and other parts of South America Bladderworts of this type. But those which we found to-day, growing out of the damp clay, were more like in habit to a delicate stalk of flax, or even a bent of grass, upright, leafless, or all but leafless, with heads of small blue or yellow flowers, and carrying, in one species, a few very minute bladders about the roots; in another none at all. A strange variation from the normal type of the family; yet not so strange, after all, as that of another variety in the high mountain woods, which, finding neither ponds to float in nor swamp to root in, has taken to lodging as a parasite among the wet moss on tree-trunks; not so strange, either, as that of yet another, which floats, but in the most unexpected spots, namely, in the water which lodges between the leaf-sheaths of the wild Pines, perched on the tree-boughs, a parasite on parasites, and sends out long runners as it grows along the bough in search of the next wild Pine and its tiny reservoirs.

"In the face of such strange facts is it very absurd to guess that these Utricularias, so like each other in their singular and highly-specialised flowers, so unlike each other in the habit of the rest of the plant, have started from some one original type, perhaps long since extinct; and that, carried by birds into quite new situations, they have adapted themselves, by natural selection, to new circumstances, changing the parts which require change—the leaves and stalks; but keeping comparatively unchanged those which needed no change—the flowers?"

Dr. Gardner mentions a Brazilian species of Bladderwort which is only to be found growing in the water collected in the bottoms of the leaves of a large *Tillandsia*, in an arid part of the mountains, at an altitude of 5000 feet. The Bladderworts seem long ago to have diverged in their habits;

the tropic species have become semi-parasitic and epiphytal, and the temperate species, chiefly *Carnivorous* in their character, although there are Indian and Malayan species of Utricularia also possessed of bladders, in which insects are caught.

The tide of vegetable life is always ebbing and flowing. Whilst some species of plants are acquiring new habits, others are forsaking their more recently formed ones, and returning to their ancestral modes of life. We have in our existing floras plants which are becoming insectivorous, both after the Sundew fashion, and also after that of the Pitcher-plants. The connate leaves of our wayside Teasel (*Dipsacus sylvestris*) may be found in the summer time with their basins filled with water, in which float the bodies of drowned ants and other insects. What is this but the foundation both of the habit and the architecture of a Pitcher-plant? The Catchflies (*Silene*) always have the upper parts of their flower-heads and stems coated with dead flies, which have been limed by the numerous sticky glands. This may be regarded as the commencement of the Sundew habit of carnivorous plants.

On the other hand, some botanists regard the exquisitely pretty Grass of Parnassus (*Parnassia palustris*) as an insectivorous flower, which has given up the habit, but which still retains in the remarkable clusters of organs seen inside the petals evidences of its former career. Dr. Müller, however, thinks these are deceptive organs, simulating the appearance of minute drops of water, to allure flies.

FIG. 100.—Grass of Parnassus (*Parnassia palustris*).

CHAPTER XIV.

GEOGRAPHICAL VICISSITUDES OF PLANTS.

ONLY those familiar with the geological history of Great Britain are aware of the numerous, extreme, and extensive geographical and climatal changes which have taken place in these latitudes since our familiar terrestial plants first appeared. With slow, but certain movements, extending over hundreds of thousands of years, pendulum-like the climate has swung from one extreme to another—from tropical heat to Arctic cold.

The period in which we live is, as regards our northern hemisphere, half-way between these extremes, or *temperate* in its climate. We are passing from the last extreme condition of Arctic cold (during the Glacial Period), slowly, to an infinitesimally increasing state of warmth, which may culminate ages hence, in a tropical warmth similar to that which geologists are perfectly assured existed in this country during the *Eocene* Period of Tertiary time, when the familiar blue clay cut into during the too frequent metropolitan excavations was being slowly accumulated on a shallow sea-floor.

For ages before that—indeed ever since the dry land received its first lowly-organised vegetation in Cambrian times—this slow beat of climatic changes has repeatedly taken place. What strange extremes the land-plants must have been subjected to! It is as if the Esquimaux had been repeatedly transported to Calcutta, and the Hindoos forced to take up their abode amidst the eternal frosts and snows of the Arctic regions!

It must be further remembered that these climatic variations have been accompanied by equally slow but certain geographical changes, which have converted dry land into sea-bed, and upheaved the bottom of the sea to form new dry lands—which have crumpled up previously horizontal strata into tablelands and mountain-chains; or depressed those already existing until they were let down in the middle like a bellying chain, as is the case with the underground connection between the Mendip and Ardennes Hills, beneath London.

We can hardly overstate the influence such climatal and physiographical changes must have had upon terrestial floras. As soon as a comparative equilibrium had been set up, and the plants had relatively fought out their battles and yielded to the strongest—as soon as all had more or less adapted themselves to the surrounding physical conditions—a change commenced, which ultimately required the whole process to be done over again. A warmth-loving flora, like that we know existed in England during the Eocene Period, had settled down pleasantly, and seemed to be irremovable. This state of things must have existed for many hundreds of thousands of years, when the climate gradually toned down, more and more, during the Miocene Period; and, when the succeeding Pliocene era set in, it inaugurated a temperate climate like that now prevailing. This was followed by the long, bitter climatal winter of "The Great Ice Age." Even if the warmth-loving Eocene plants had been able to adapt themselves to the lower temperature, they could not have lived in Great Britain during the latter period, for during one stage, at least, of that bitter time, these islands were swathed in an ice-sheet just as Greenland is now, and this would effectually prevent plants from maintaining their ancient habitats.

The case must have gone equally bad with them, even if the climate remained practically unchanged, when one of the numerous oscillations of level occurred which lowered the dry land until it became the bed of a sea.

In both these instances there would be only one way in which plants could perpetuate their species when their ancient sites had been taken from them —they would have to *migrate*. The British plants of the Eocene Period, under stress of altered climate, had moved farther *south*, and we accordingly find them now forming part of the flora of South Africa, and Brazil, etc., such as the *Proteaceæ*—the flora occupying the low lands, which were first submerged during a change of level, would be forced to retire farther inland, and grow at higher elevations. In both such cases as these there would be a tendency to overcrowd the surface of the dry land, and to cause a keener struggle for existence, resulting in the extinction of some kinds, and the complete alteration of others, under the influence of natural selection.

Large islands and parts of continental areas have been broken up, by changes of level, into small islands. These would still retain some of the old plants of the mainland, whose altered surroundings would have to be responded to in changed and adapted structures.

Suppose such a geological change to take place when the climate was very warm—all the plants on the newly-formed islands would then necessarily be warmth-loving. If a climatal change occurred, the increased cold would not go so hardly with the island plants as with those of the adjacent mainland, because an island climate is always more regular and warmer than the continental climate in the same latitude. So that whilst the plants on the latter would change, or their places would be taken by cold-loving plants from higher latitudes, those on the islands would be better circumstanced, and also free from invasion. Now let ages elapse, and an upward movement once more connect the island with the mainland, so that it formed an integral part of it—we should then have the phenomenon of two distinct floras living side by side.

Or let the opposite fact be imagined. Islands are formed during a Glacial Period, and the characteristic flora remains with them. The climate alters; some of the more sensitive species die out (although plants used to cold adapt themselves sooner to increased warmth than those which love heat do to cold); others creep farther up the hill-sides to the summits. When the intervening sea-bed is upheaved, these hilltops will have been raised still higher above the sea-level and into a colder region of the atmosphere. Or suppose a country to be very cold and occupied by an Arctic flora, and its geographical connections such that it is part and parcel of an extensive continent, or series of continents (like the land-masses of the northern hemisphere). Then, if the climate slowly changes to warmer conditions, as the snows cease to lie permanently on the mountain-slopes, and to fill up the valleys with their glaciers, the cold-loving plants of the plains, finding the climate changing, would creep up to higher and therefore colder levels, and occupy the ground previously covered with snow and ice. Eventually the tops of all the high hills and mountains would be permanently covered with what had been an *Arctic* flora of the plains, now converted into an *Alpine* flora. The ancient habitats down below would gradually be occupied by plants better fitted for the changed warmer conditions. They would swarm in like a vegetable tide, to relieve the overcrowding of their metropolitan areas. Such a country might then be subjected to a geological alteration of level, and be completely surrounded by the sea. In which case its mountain sides and hill tops would be occupied by growing Alpine plants of the same species as those found living at the sea-level in higher latitudes, and its plains would be crowded by new-comers which disliked extreme cold, and which had streamed in from quite a different quarter to that whence the Arctic flora originally started on a similar invasion, to oust warmth-loving plants of a higher antiquity still!

This need not be regarded as a series of mere suppositions. It is a brief but definite outline of the geological and climatal changes through which our English flowering plants have passed since the commencement of the Miocene Period. There are numerous plants, with which most ordinary botanists are familiar, which illustrate both the later geological periods and each of the later geographical changes as well.

Professor Edward Forbes, long ago, showed that our British flora had come to us from several points of the compass. The Marsh Wood Spurge (*Euphorbia pilosa*), found in Somersetshire, is an outlying member of the flora which existed here during the Pliocene Period, before the Irish Sea, the English Channel, and the Bristol Channel had been formed, and when the British Islands were a western extension of the European continent. At that time one could have wandered uninterruptedly from England to France and Spain, for the land extended much farther into the Atlantic than it does now. This Marsh Wood Spurge is a common Mediterranean species. The Large-flowered Butterwort (*Pinguicula grandiflora*), found only in Ireland, and the rarer Pale Butterwort (*Pinguicula lusitanica*), met with only in a few places in the west of England, are characteristic Spanish plants, as is also the Fringed Rock Cress (*Arabis ciliata*) gathered near the sea in the south of Ireland. The latter part of the British Isles, as well as tracts in Devonshire, Cornwall, Somersetshire, etc., possess many similar plants, whose characters are decidedly *southern*, and whose headquarters are in the south of France, Spain, Portugal, and along the Mediterranean coasts. It was to account for these and other anomalous groups of plants in our British flora that Professor Edward Forbes, in his *Geological Relations of the Fauna and Flora of the British Isles*, thirty years ago suggested a former land union between England and the Continent by way of the English Channel; and this brilliant theory has since been verified in several ways, so that geologists have now no doubt such a terrestrial connection did exist during the Pliocene Period.

To look over a list of British plants, drawn up in botanical order, is to the botanist, acquainted with the various origin of our flora, an occupation of great interest. He there finds arranged in categorical order species which came from centres geographically wide apart, and whose habits are of the extremest and most opposite character. Here for instance, is the Alpine Butterwort (*Pinguicula alpina*) side by side with the Pale Butterwort (*Pinguicula lusitanica*) above mentioned. It is certain that the former has wandered down here from the Arctic regions, and we are not less sure that the latter has crept up from the Mediterranean area, or remained behind ever since the early Pliocene Period.

During the Pliocene Period, the dry land was occupied, perhaps, by the same flora as we now find scattered over France, Spain, and Portugal. The North Sea separated England from what are now Norway, Holland, and Belgium, and came down as far as Kent. Beyond that the chalk extended in an unbroken succession to France, for the Straits of Dover were not formed until a much later period.

As the temperature changed and became colder, one by one these warmth-loving plants died out, until eventually only a few were left in the southern parts of the British Islands. We find them most abundantly there, because the ice-sheet of the Glacial

Period did not extend southward much, if any, beyond what is now the Thames; and also because they find in the mild winters of Cornwall, Devon, and the Channel Islands, a climate similar to that they originally enjoyed, and under whose influence they were perhaps evolved.

Therefore these *southern* plants belong to an earlier period. They are "ancient Britons" by comparison with the *northern* or Arctic plants; and even the latter are of much higher antiquity than the main flora of our fields, meadows, and green lanes, which did not occupy England until the Glacial Period had passed away, with its severe cold, and the present climate had been instituted. The area now occupied by the German Ocean was then dry land, through which the river Thames made its way, to be eventually joined by the Rhine, into the North Sea off the Scottish coasts. It was by way of this dry-land connection with Holland and Belgium that our lowland flora reached us from Germany. So that the main army of our common flowering plants are new-comers. They occupy the richest soils, whence they ousted the older and less fitly adapted species—just as the Normans possessed themselves of Saxon lands, and the Saxons of the earlier British possessions. It is on our hills and mountains that we find the older Arctic flora of the Glacial Period protected from invasion; precisely in the same spots that the remnants of the ancient

British tribes took shelter, as the Welsh and Gaelic languages of North Wales and the Highlands of Scotland indicate.

As might be expected, when we consider all the circumstances, our southern flora is by no means so numerously represented as the northern. The latter may be found on all our hills and mountains, marvellously increasing in number of species as we get farther on into Sutherlandshire, as was to be looked for, although it must not be forgotten that the sea cuts it off now from its original Scandinavian home.

During the extreme cold of the "Great Ice Age," this Arctic flora extended as far south as the Pyrenees. Hence we find there many of the same kinds of plants as now grow within the polar circle. The Alps of Switzerland are equally rich in Arctic species, and many of the most beautiful kinds collected by the tourist, high up the mountains, belong to this group. Out of six hundred and eighty-five species of flowering plants now found in Lapland, one hundred and eight grow in Switzerland, and sixty-eight in the Pyrenees.

Darwin makes especial mention of these ancient floral migrations, and in one place, after describing the close of the Glacial Period, he says: "As the warmth returned, the Arctic forms would retreat northwards, closely followed up in their retreat by the productions of the more temperate regions, and as the snow melted from the bases of the mountains the Arctic forms would seize on the cleared and thawed ground, always ascending, as the warmth increased and the snow still further disappeared, higher and higher, whilst their brethren were pursuing their northern journey. Hence, when the warmth had fully returned, the sames species which had lately lived together on the European and North American lowlands, would again be found in the Arctic regions of the Old and New Worlds, and on many isolated mountain-summits far distant from each other. Thus we can understand the identity of many plants at points so immensely remote as the mountains of the United States and those of Europe. We can thus also understand the fact that the Alpine plants of each mountain range are more especially related to the Arctic forms living due north, or nearly due north of them; for the first migration when the cold came on, and the re-migration on the returning warmth, would generally have been due south and north. The Alpine plants, for example, of Scotland, as remarked by Mr. H. C. Watson, and those of the Pyrenees, as remarked by Ramond, are more especially allied to the plants of northern Scandinavia; those of the United States to Labrador; and those of the mountains of Siberia to the Arctic regions of that country."

The floras of many countries have been, after the fashion of serial books, "taken in in numbers," and bound afterwards; but that of Great Britain is of an especially fragmentary character. Still, both in its general make-up and dissimilar origin, it is very nearly related to that of the mainland of Europe.

Of the British flora in pre-Tertiary times, its origin and geographical migrations, we know very little, except that during the Oolitic and Wealden Periods the dry land was certainly richly clad with *Zamias*, and other Cycadaceous plants, as well as with peculiar kinds of Coniferæ. In the mid-Secondary epoch a great northern continent, uncovered, perhaps, with ice and snow, stretched to the north pole. It was watered by mighty rivers, vast as those now peculiar to the tropics, some of which converged, and when they poured their waters into the sea formed banks of sedimentary material which ultimately grew into a delta. The Wealden formation represents such an ancient delta in British geology. It formerly extended in unbroken continuity, after its upheaval, to France, Belgium, and even into Germany, but it has been severed by the denudation of the English Channel. This Wealden formation in places contains abundant relics of terrestrial vegetation; and, what is still more to the point, as attesting the luxuriant vegetation which the continental conditions and warmer climate may have spread to the pole itself, there are remains of *Iguanodons* and other remarkable herbivorous reptiles, whose presence indicates abundance of vegetation and forest conditions.

It is doubtful whether our British flora contains a solitary member which was then living, unless they are represented by Club-mosses, Horsetails, and Ferns. Our common Bracken (*Pteris aquilina*) may have been a spectator; and the Cycadaceæ of the Cape, the Brazils, and Australia, are undoubtedly lineal descendants of the forms which grew so abundantly in these latitudes during the early Oolitic Period.

Whatever may have been the composition of the terrestrial flora in what is now England, during the age just mentioned, it must all have been swept away by the great change which took place during the latter part of the Cretaceous Period, for all the dry land was submerged, and a tolerably deep sea eventually covered it, along whose bottom the "white chalk" of England was slowly elaborated. The chalk has been eaten away on every side since it was upheaved and converted into dry land, but it still extends from Ireland to Bordeaux, and from Sweden to the Crimea—an area which will give some idea of the extensive Cretaceous ocean, and which will convey to the mind what a large tract of dry land, with its crowds of plants, must have been submerged before that Cretaceous ocean could have been formed.

Flowering plants of familiar types, allied to our Oaks, Maples, etc., had appeared upon the earth before the Cretaceous ocean was in actual existence;

and we cannot suppose they were the only kinds of flowers. Hosts of herbaceous plants less capable of being preserved must have been their compeers. In short, the battle of life was being fought out as keenly and bitterly in that distant epoch as it is now, and perhaps as many individuals, if not as many *species* of plants, joined in the fray.

When we fully bear in mind that flowering plants have been subjected to these multitudinous external changes, and that their living tissues are so plastic that they can adapt themselves, unless overstrained, to a large range of variation, the wonder is that modern types retain so much of the character which distinguished them ages ago—that Oaks should still be Oaks, and Maples, Maples, now as they were in pre-Cretaceous times.

Notwithstanding the numerous evidences of floral change which have already been noticed, the flowering parts of plants are not subjected to such changeable conditions as roots and stems. These have to be adapted to a variety of physical and chemical conditions—to dry and moist places, and to soils which may contain this or that chemical element. Hence we cannot wonder that in the same genus of plants the flowers are remarkably alike, differing only in some unimportant features, easily explained, of size and tint of corolla, whilst the leaves run the entire gamut of foliar pattern—as, for instance, is the case with our Buttercups. This remarkable variation is the measurement of the response the plants have made to the different physical conditions which set in, it may be, ages ago, and which have determined the habitats of the species ever since. Not unfrequently the type of the primitive leaf of a genus (and perhaps even of an *order* of plants) will be revealed in the shapes of the *radical* leaves, which are nearly always different from the later-formed leaves. In the different species of Buttercups these radical leaves are remarkably alike, and in striking contrast to the variable patterns of the leaves characteristic of different species.

In the northern hemisphere, and for the matter of that in the southern hemisphere also (although the two periods may not have been contemporaneous, but successive), the most remarkable of all the geological epochs, with the exception of the Miocene, for abounding vegetation, was the Carboniferous. But if we compare the flora of these two widely-separated epochs we cannot help being struck with their contrast. The Carboniferous flora abounded in individuals. Its character was extremely monotonous. It is doubtful whether true flowers (with the exception of those of the Coniferæ) had then appeared, for *Antholites* is now degraded from its supposed rank as a fossil Aroid to a species of Horsetail. Palms, once supposed to have been present in some abundance, are now known to have been few, and some botanists doubt the genuineness of the remainder. A few cycadaceous plants may have commenced existence. On the other hand, the flowerless plants (*Cryptogamia*) had attained their maximum of differentiation, adaptation, magnitude, and geographical distribution. Perhaps every one of the existing types of cryptogamic vegetation, except Mosses and Lichens, had been evolved at that distant period, not even excepting the Rhizocarps. Since then these types have been more or less stereotyped; Tree-ferns, Club-mosses, Horsetails, and Rhizocarps have not been materially modified, except in size. Few of them have been able to maintain their Carboniferous bulk, for they have had to contend with new-comers of higher organisation.

Similar characters distinguish the Carboniferous floras of all countries—Europe, America, China, and Australia.

The insects found fossilised in the Coal formation are similarly scanty, and limited to one or two orders. This general absence of flowers and insects in the Carboniferous strata is very suggestive.

Millions of years must have passed away, from the time when our coal-beds were formed, through excess of vegetable life, to the Miocene Period. We get numerous glimpses of the progress made by the vegetable kingdom in the meantime, one of which we have glanced at during the Wealden. But the Miocene flora bursts upon us in magnificent development and modification. Its most striking peculiarity is its generalised geographical character. In Europe there were gathered together types of plants, all growing side by side, which are now only to be found in America, Japan, China, the Cape of Good Hope, and even Australia; in addition to those which lingered on, and of which the Mediterranean and North African species are the lineal descendants. Had the Miocene vegetation been a literal "Garden of Eden," and its inhabitants been subsequently "driven forth," they could not have been dispersed more effectually. What a silent but effective story is associated with the dispersal of such genera, for instance, as *Sequoia* (*Wellingtonia*), once abundant in Europe, now confined to solitary Californian valleys! The extent of this dispersal of the Miocene flora is a measure of the great physical changes which have taken place since they all grew together.

Not only are the Miocene plants abundant in orders and genera, but those in which floral differentiation and high organisation have been evolved—as the *Compositæ*—are represented, showing plainly that all the leading botanical features had then been developed.

Nor can we wonder, for the number of species of fossil plants recorded is only equalled by the immense number of fossil insects of all kinds—from ants, bees, etc., to beetles,—all associated with the evolution of flowers. The remains of fossil birds—Trogons, Parrakeets, etc., equally indicate the presence of creatures associated with the development of coloured and pulpy fruits ; whilst the bones of fossil monkeys, rodents, etc., suggest that nuts, with their hard coverings, must already have been formed, by way of defence against decimation by such mammalia.

In short, all the principal features of the vegetable kingdom, such as we now behold, had been more than roughed out—they had been refined—in the Miocene Period. The chief fact in connection with flowering plants since then has been their altered geographical distribution, due both to change of climate and the physical geographical changes which have occurred in the meantime. Under the influence of these combined, many species and some genera have become extinct. Others which were then abundant are now rare, and contrariwise. Habits of parasitism may have been developed on the part of members of various orders—perhaps insectivorous habits as well. Some plants which formerly possessed upright stems may now be climbers or creepers ; some may have developed thorns and prickles out of modified leaves, stipules, branches, or hairs, in response to the greater dangers surrounding them from the introduction of greater numbers of mammalia. The flowers of many orders may have been changed in colour and even shape, so as to produce the extreme varieties we have in such orders as *Ranunculaceæ*, *Rosaceæ*, *Scrophulariaceæ*, etc.

The process is still going on, and must go on. Organic nature knows no more of rest or stability than does the surface of the ocean. The vegetable kingdom has had its ebbs and flows, its developments and degenerations. The story of the life of plants is one nearly analogous to that of the life of men—full of failures and successes, of ambitious efforts, and degenerations ending in extinction! The path of human history is thickly strewn with projects and follies, and the strata of the earth's crust are equally abundant in the remains of fossil plants belonging to genera no longer existing, showing us that the pedigree of the vegetable kingdom has suffered the same incidents as that of humanity. In each case out of weakness has been evolved strength. The extinct types of plants served as a platform on which more highly-organised forms could be evolved—the follies of mankind have been the means of developing higher wisdom and prudence. Even degeneration, both in plants and men, has originated new characters and unexpected habits.

To myself it does not seem possible to contemplate this tangle of Vegetable Life and its conditions, geological, astronomical, geographical, and biological —this series of progressions, degenerations, modifications, and adaptations, rising and falling during the millions of years which have passed away since it began, and which has culminated in the wonderful flora now possessing the earth—without feeling that underneath and behind all are the Untiring purposes of Divine Wisdom and Love!

INDEX.

ABORTED ORGANS, 193.
Acacias, tannin in, 129.
Acer campestre, 71.
Aconitum, 122.
Acorns, cotyledons of, 18.
Adventitious buds, 57, 59.
Agaric, fly, 126.
Air-plant, 57.
Albuminoid substances of plants, 32.
Algæ, zoospores of, 3.
Allen, Grant, 120, 124, 139.
Aloe, flowering of, 14.
Alpine Gentians, flowering of, 15; flowers, 161; and butterflies, 165.
Amœba, the, 3.
Ampelopsis hederacea, 54.
Anemone ferox, 122.
Anemophilous flower, 34, 67, 72.
Anona muricata, 242.
Antherozoa, 3.
Antholites, 299.
Anthrophora, 82.
Antiquity of certain flowers, 60.
Apocynum, 89.
Apple blossom, 173.
Apples, 79.
Apricots, 95.
Arabis ciliata, 291.
Arctium lappa, leaves of, 107.
Armstrong on red clover, 223.
Artichoke, the, 211.

Arum maculatum, 104, 120, 124, 139.
Asa Gray, 90, 271.
Asclepiadaceæ, the, 89.
Ash, the, 91; samara of, 112.
Atropa belladonna, 104, 118.

BACTERIA, the, 115.
Balsam, the, 109.
Bates, Henry, 228, 231.
Bauhinias, the, 48.
Bee Orchis, 6, 68.
Belt, Thomas, 91, 143, 154.
Bennett, A. W., 137.
Bignonia capreolata, 54.
Bird-disseminated fruits, 103.
Birds and fruit, 94; and poisonous plants, 120.
Birthwort, 77, 139.
Bitter-sweet, 53, 104, 118.
Blackberry, the, 90.
"Black-mail" paid by plants, 143.
Bladderworts, the, 279, 280, 281; Kingsley on West Indian, 281.
Blind flowers, 70.
Bloom on fruit, 121.
Bog-bean, the, 127.
Boraginaceæ, double colours in, 82, 173.
Bougainvilleas, 202.
Bracken, the, 47, 129.
Brambles, the, 98.
Brazil nuts, 134.

British Spurges, 200.
Broad bean, embryo of, 19, 20.
Broom-rapes, the, 249.
Bryony, 104.
Bryophyllums, the, 57, 58, 59.
Buckwheat, 53.
Bulbils, 190.
Bulbs and mammalia, 127.
Bull's-horn Thorn, Acacia, 144.
Bunbury, Sir Charles, 151, 265.
Buphane toxicaria, 127.
Burdock, the, 107.
Bush-ropes, 47.
Buttercups, secretion of, 123.
Butterworts, 127, 279.

CACTUS, leaves of, 205.
Cactuses and *Euphorbias*, 151.
Caladium seguinum, 123.
Calamites and *Equisetaceæ*, 130.
Caltha palustris, 196.
Campanulas, the, 77, 81.
Campion, night-flowering, 82.
Cannon Ball-tree, 135.
Carboniferous flora, the, 299, 300.
Cardamine pratensis, 59.
Carline Thistle, 111.
Cassytha filiformis, 248.
Catchflies, the, 142.
Cattle and poisonous plants, 49.
Celandine, the, 84.
Cells, shapes of, 22.
Centranthus ruber, 112.
Cephalotus, 278.
Cestrum, 118.
Cherry, 94, 99.
Chlora perfoliata, 6, 127.
Chlorophyll of cells, 23, 28.
Circulation of sap, 26.
Citric acid in fruits, 126.
Classification of Linnæus, 33, 61, 70.
Cleavers, the, 47, 50, 107.
Cleistogamic flowers, 219; Darwin on, 222; and ovaries, 224.
Clematis vitalba, 53.

Climbing plants, European, 237; Fumitory, 53.
Clock of Dandelion, 110; Goat's-beard, 110.
Club-mosses, ancient, 43.
Coloured calyces of flowers, 195.
Colours, double, of flowers, 82.
Commensalism, vegetable, 227.
Compositæ, seed dissemination of, 109.
Cooke, Dr., 108, 109, 277.
Co-operation in flowers, 167; fruits, 177; perfumes, 177.
Cope, Prof., on volition, 9.
Corydalis claviculata, 53.
Cotton-plant, 112; grass, 112.
Couch-grass, 13.
Creeping stems of strawberry, etc., 190.
Crepis pulchra, 141.
Cretaceous period, flora of, 299, 300.
Cruciferæ, 124.
Cuckoo-flower, buds of, 59.
Cucumber, Squirting, 108.
Cuscuta Europæa, 246.

DANDELION, pappus of, 110.
Darlingtonia, the, 274.
Darwin, 1, 53, 67, 71, 80, 87, 88, 90, 222, 225, 268.
Dewberry, the, 98.
Dicotyledons, 19.
Digitalis purpurea, 124.
Diœcious flowers, 92.
Dionæa muscipula, 266.
Dipsaceæ, 112, 177.
Distribution of Sundews, 266.
Dodders, the, 245.
Dog's Mercury, the, 74.
Drooping flowers, 195.
Drosera rotundifolia, 260.
Dumb-cane, the, 130.
Duration of plant life, 214.
Dutch Rushes, 130.

ECHIUM VULGARE, 82.
Elm, 59, 71, 112.

INDEX. 307

Embryo of bean, 19.
Entomophilous flowers, 34, 67, 68.
Eocene period, flora of, 287.
Epiphytal Orchids, 206; plants, 240.
Equisetums, 130.
Ericaceæ, the, 102.
Eriophorum, 112.
Erodium, 193.
Erythrium centaureum, 126.
Euonymus europæus, 72, 105.
Euphorbiaceæ, 117, 200.
Euphrasia officinalis, 252; *pilosa*, 291.
Evening Primrose, the, 81.

FERNS, tannin in, 129; leaves of, 192.
Ficus repens, 49.
Fig, receptacle of, 100.
Figwort, the, 77, 124.
Fir, 103.
Fission and Air-plant, 57.
Flora, British bird, 103.
Floral altruism, 168.
Floral poverty, 210.
Flowers, nature of, 11; of Aloe, 14; Alpine Gentians, 15; of Nettles, 33; of Common Vine, 33; anemophilous, 34, 72; entomophilous, 34; ovary of, 36; blind, 70; hermaphrodite, 71; diœcious, 72; monœcious, 72; proterogynous, 77; proterandrous, 77; colours of, 81; white, etc., 81; double colours in, 82; gamopetalous, 81, 83; irregular, 85; and humming-birds, etc., 90; and fruits of Horse-chestnut, 211.
Flowering, Herbert Spencer on, 60; of grasses, 74.
Food-store of seeds, 180; bulbs, 183; roots, 188.
Forbes, Prof., on origin of British flora, 280, 290, 291.

Forget-me-not, the, 82.
Fossil funguses, 252.
Foxglove, the, 124.
Fraxinus excelsior, 71.
Fruits, Ruskin on, 93; and birds, 94; and men, 95; average sizes of, 96; hard stones of, 97; colours of, 102; bird-disseminated, 103; nauseous, 104; bloom of, 121; acids in, 126.
Functions of stomata, 27.

GALIUM APARINE, 47, 107.
Gamopetalous flowers, 81, 85.
Gentianaceæ, the, 126.
Gentians, 77.
Gentians, Alpine, 15.
Geraniaceæ, the, 193.
Geraniums, 77.
Glacial period, the, and plants, 285; flora of, 287, 289, 293, 294; Darwin on, 294.
Glumes of Grasses, 132.
Goat's-beard, 110, 140.
Gonophores of *Sertularia*, 12.
Goose-grass, the, 47, 107.
Gorse leaves, 152.
Gorse, seeds of, 109.
Gosse, Mr., 122, 221, 233.
Gossypium, 112.
Grant Allen, 31, 81, 85, 107, 120, 134, 154, 199, 211.
Grasses, flowering of, 74.
Gray, Dr. Asa, 271.
Growth of plants, 8, 20.

HAMMELIS VIRGINICA, 108.
Harpagophytum, 108.
Harveya, 251.
Hawthorn, the, 84.
Hazel, 94; American Witch, 108.
Heath, Common, 129.
Hibbertia, 53.
Hippomane mancinella, 104.
Hips, 100.
Holly leaves, 154.

Honey glands, 33.
Honeysuckle, 104, 127.
Hooker, Sir Joseph, 277.
Hop, 74.
Hop, hooks of, 51.
Horse-chestnut, 71.
Horsetails, 130.
Hoya carnosa, 49.
Humboldt, 44.
Humming-birds and flowers, 90.
Hydra, 58.

IMPATIENS-NOLI-ME-TANGERE, 109.
Insect crossing, 34; protections to plants, 146.
Insectivorous plants, 258.
Iris fœtidissima, 105.
Ivy, growth of, 49.

JUNIPER, 103, 104.

KERNER, 136, 140, 142, 143, 158.
Kingsley, Charles, 135, 195, 227, 233, 235, 264, 281.
Knapweeds, the, 170.

LABIATÆ, the, 85.
Labiums of flowers, 85.
Land, sinking of, 10.
Larvæ of beetles, 125.
Lathræa squamaria, 251.
Lauder Lindsay, Dr., 237.
Leaf-buds, 175; bracts, 176.
Leaves of Coniferæ, 131; shapes of, 31.
Lichens, 44, 237.
Life-history of Aphis, 12.
Lignine in cells, etc., 23.
Lime, the, 113.
Lindley, Dr., 113.
Linnean system of classification, 33, 61, 70.
Liver-coloured flowers and flies, 138.
Lodicules of Grasses, etc., 209.
Loosestrife, the, 80.

Lophospermum, 54.
Loranthaceæ, the, 242.
Loranthus Europæus, 243; *glaucus*, 243.
Lords and Ladies, the, 120.
Lowland floras, 293.
Lubbock, Sir John, 89, 137, 140, 141.
Lychnis vespertina, 82.
Lycopersicum esculentum, 118.
Lygodium, twining habits of, 52.
Lythrum salicaria, 80.

MACLURA AURANTIACA, 57.
Malic, etc., acids in fruits, 128.
Mallow, 77.
Manchineel, the, 104, 117.
Manihot, 117.
Mankind and fruits, 95.
Maple, 57, 71, 112.
Marcgravia, 47, 91.
Matapolo, the, 236.
Menyanthes trifoliata, 127
Micropyle of ovule, 36.
Milkweeds, the, 89.
Milky juices of plants, 158.
Mimicry in fruits, 133.
Miocene period, flora of, 290, 300; plants, distribution of, 301, 302.
Mistletoe, the, 102, 238, 239, 241.
Modes of inflorescence, 166.
Modified leaves, 182, 203; leaf-stalk, 203.
Momordica elaterium, 108.
Monkey-pot tree, the, 135.
Monocotyledonous seeds, 19.
Monœcious flowers, 72.
Monstrosities in flowers, 12.
Montpellier, foreign seeds at, 108.
Moths, 81.
Mountain Ash, the, 99.
Mullein, the, 124.
Müller, Dr., 81, 82, 85, 90, 91.
Murderer Liana, 229.
Myosotis, 82.
Myzodendron, 244.

INDEX. 309

NEPENTHES, the, 257, 274.
Nervous structures, 3.
Nettle, flowers of, 33.
Nightshade, Deadly, 104; Black, 104, 118.
Nipplewort, 141.
Nodes, suppression of, 30.
Nottingham Catch-fly, 252.
Nux-vomica, seeds of, 119.

OLIVER, Prof., 122.
Oolitic flora, 296.
Ophioglossum japonicum, 52.
Ophrys apifera, 6.
Orchidaceæ, the, 85.
Orchids, 87, 88.
Orchis, the Bee, 6.
Orobanche rapum, 252.
Ovary of flowers, 36.
Ovule, micropyle of, 36.
Oxalic acid in plants, 122.
Oxalis acetosella, 126.

PÆONY, pollen of, 221.
Palms, pollen of, 221.
Pappus of Thistle, 102; Dandelion, 109; Goat's-beard, 110; Ragwort, 111; Teazle, 112; *Centranthus*, 112.
Parasitic Fungi, 254.
Parnassus, Grass of, 283.
Passiflora gracilis, 54.
Peach, 94.
Pears, 99.
Perfumes, protective, 137.
Periploca Græca, 53.
Petaloidea, the, 194.
Phyllodes, 203.
Phyllotaxy, 30.
Pine, 45, 102.
Pines, pollen of, 221.
Pin-eyed primroses, 79.
Pinguicula vulgaris, 127; *grandiflora*, 291; *alpina*, 292; *lusitanica*, 292.
Pinus pumilio, 133.
Pistil of Vine, 33.

Pitcher-plants, the, 257, 274, 275.
Pithecolobium, 106.
Plants, instincts of, 4; ringing, 11; colony of leaves, 13; functions of, 24; root-sheaths of, 25; protoplasm in, 31; albuminoid substances in, 32; poisonous nature of, 117, 119; silica in, 130.
Poison berries, 118; bulbs, 127.
Pollen-grains, 33, 35.
Pollinia of Orchids, 88.
Polygonum convolvulus, 53; *amphibium*, 142.
Polypetalous flowers, 81.
Poplar, the, 112.
Potato, the, 56, 118.
Prickles of Brambles, 150.
Pringlea, the, 68.
Proboscidea jussieui, 108.
Proteaceæ, the, 288.
Protection against Aphides, 147; of thorns and prickles, 149.
Proterandrous flowers, 77.
Proterogynous flowers, 77.
Protoplasm in plants, 31.
Protozoa, the, 3.
Psychology, vegetable, 5.
Pteris aquilina, 47, 129, 297.
Puccinia graminis, 255.

RADICLE of bean, 20.
Ragwort, 111.
Ranunculaceæ, 122, 123, 126, 196, 197.
Ranunculus ficaria, 84; *acris*, 123.
Raspberry, 98.
Ray Lankester, Prof., 207.
Razor-grass, 235.
Red Clover, 223.
Reproduction in plants, 32.
Rhinanthus crista-galli, 252.
Rhizomes of plants, 189.
Rhododendrons, 41.
Riley, Dr., on *Sarracenias*, 273.

Ringing plants, 11.
Roast-beef plant, 105.
Roman tombs, seeds in, 113.
Root tips, 1; sap of, 23; sheath, 25.
Rosaceæ, variation in the, 252.
Roses, 98.
Rumex acetosa, 126.
Ruskin, Prof., on fruits, 93.

SALICACEÆ, variation in, 216.
Saltwort, 154.
Salvias, 87.
Samaras of Maple, etc., 112.
Sanderson, Dr. Burdon, 268.
Sap of roots, 23; circulation of, 26.
Sarcode, comparison of, with sap, 12.
Sarracenias, the, 269, 273.
Scrophulariaceæ, the, 85, 124.
Sea-fir, the, 12.
Seeds, 108, 113.
Self-fertilisation, Darwin on, 67.
Senecio jacobæa, 111.
Service-tree, the, 99, 101.
Shapes of leaves, Grant Allen on, 31.
Sickening of land, 10.
Silica in plants, 130, 133.
Sizes of fruits, 96.
Smilax, 51.
Snakes and poisonous plants, 119.
Social plants, 161, 163, 165.
Solanaceæ, poisonous character of, 118.
Solanum dulcamara, 53, 118; *nigrum*, 118; *tuberosum*, 118.
Sorrel, Common, 126; Wood, 126.
Spanish chestnut, pollen of, 221.
Speedwells, the, 124.
Spencer, Herbert, 58, 60.
Spindle-tree, the, 105, 106.
Spines of Thistle, etc., 153.
Sports in plants, 12.
Spring flowers, 162.
Stamens, 53.
Stapelias, the, 138.

Stems, nodes and internodes of, 30.
Stomata, function of, 27.
Stones of fruits, 97.
Strawberry-roots of, 58; common, Darwin on, 71; common, 99; receptacle of, 99.
Strychnos genus, the, 119.
Strychnos nux-vomica, 119.
Sumachs, the, 117.
Sundews, the, 258.
Suppression of nodes, etc., 30; of floral parts, 210, 211.

TAMUS ELEPHANTIPES, 53.
Tannin in bark, 121, 128; fruits, 128; ferns, 129.
Teazel, 112, 127, 283.
Tecoma radicans, 50.
Tendrils, 51; sensitiveness of, 54.
Tentacles of Sundews, 262.
Teratology of Cabbage-leaf, 269.
Thalictrum, stamens of, 197.
Thistle, 99, 102, 110.
Thrum-eyed Primrose, 79.
Thymus serpyllifolia, 72.
Tilia Europæa, 113.
Toad-stools and beetles, 125.
Tomato, the, 118.
Tragopogon pratensis, 110.
Traveller's-joy, 53.
Treat, Mrs., on Bladderworts, 258, 280.
Tree-ferns, 44, 45.
Trifoliate leaves of Clover, 157.
Trimorphism, 80.
Tropæolum, the, 54.
Tropical Forests, Wallace on, 231; Bates on, 232; Gosse on, 233; Kingsley on, 233.
Twining of plants, 51, 52; double, of *Hibbertia*, 53.
Tyndall, Prof., on perfumes, 137.
Typical numbers in flowers, 209.

UMBELLIFEROUS FLOWERS, 171.
Utricularias, the, 279.

VALERIANACEÆ, the, 112.
Variation of plant life, 214; in sizes of plants, 215
Vegetable communism, 149; psychology, 5.
Vegetative fronds of ferns, etc., 192.
Venus' Fly-trap, the, 257, 266, 267, 268.
Verbascum, 124.
Vernal floras, 162.
Veronicas, astringency of, 124.
Vine Bamboo, 233.
Vine, flowers of, 33.
Viper's Bugloss, the, 82.
Virginia Creeper, the, 54.
Viscid secretions of plants, 142.
Volition, Prof. Cope on, 7.

WALLACE, A. R., 48, 231, 275.
Water in plants, 24.

Watson, H. C., on Scottish Alpine plants, 295.
Wealden, flora of, 296.
Wheat, pollen of, 221.
White flowers and moths, 82.
Willow-herbs, the, 77.
Willows, stamens of, 197.
Wind-crossing, 35.
Wolf's-bane, the, 122.
Wood-rushes, 209; Sorrel, pollen of, 221.
Wourah, poison of, 119.

XANTHIUMS, the, 108.

YELLOW-WORT, the, 6, 126.
Yew, the, 103, 104.

ZEA MAYS, 71.
Zoospores of Algæ, 3, 93.

THE END.

Printed by R. & R. CLARK, *Edinburgh*.

www.ingramcontent.com/pod-product-compliance
Lightning Source LLC
Chambersburg PA
CBHW030750230426
43667CB00007B/915